AF517405

SPICE CROPS PRODUCTION TECHNOLOGY

P.C. Das

SCIENTIFIC
PUBLISHERS (INDIA)

Published by:

Scientific Publishers (India)
5 A, New Pali Road, P.O. Box 91
Jodhpur 342 001 (India)

Branch Office
Scientific Publishers (India)
4806/24, Ansari Road, Daryaganj
New Delhi - 110 002 (India)

E-mail: info@scientificpub.com
Website: www.scientificpub.com

Print : 2018

© 2014 P.C. Das

All rights reserved. No part of this publication or the information contained herein may be reproduced, adapted, abridged, translated, stored in a retrieval system, computer system, photographic or other systems or transmitted in any form or by any means, electronic, mechanical, by photocopying, recording or otherwise, without written prior permission from the author and the publishers.

Disclaimer: Whereas every effort has been made to avoid errors and omissions, this publication is being sold on the understanding that neither the author nor the publishers nor the printers would be liable in any manner to any person either for an error or for an omission in this publication, or for any action to be taken on the basis of this work. Any inadvertent discrepancy noted may be brought to the attention of the publishers, for rectifying it in future editions, if published.

ISBN: 978-81-7233-873-2

Typeset by Rajesh Ojha
Printed in India

PREFECE

Spices are very important as food and as medicine and they bring out the unique natural taste of cuisines and could be used to change the look of the food to make it more attractive in colour. Keeping in mind the importance of Spice Crops, 'Spice Crops Production Technology' is published to provide the information about cultivation method along with their uses to the Students, Teachers, Extension officer and the person engaged in production of Spice crops. The book has been written in simple and lucid language under Indian context. In this book, I do not pretend to claim credit for any original contribution on the subject. Rather; it should be viewed as a concise collection of various topics from different, books, bulletins, scientific paper and wave sight etc, the names of which are mentioned in the text itself or in the selected bibliography.

I am thankful to my school mates Dr. Basudeb De, Former Associate Professor in Chemistry, Sambhunath College, Luvpur, Birbhum and West Bengal for providing me books at his disposal and by borrowing books from 'Hrishekesh Smriti Pathagar', established by himself in the memory of his father, as per my requirement. I am thankful to Dr. R.K. Ghosh, Professor of Agronomy, Dr. Debesh Pal, Associate Professor of Agronomy, Dr. A. Zaman, Professor of Agronomy and Dr. Koushik Brahmachari, Professor of Agronomy, Bidhan Chandra Krishi Viswavidhyalaya, Nadia, West Bengal and Dr. Dhiman Mukherjee, Assistant Professor of Agronomy, Uttar Banga Krishi Viswavidhyalaya, Coochbehar, West Bengal for their help.

I am also thankful to my life partner Mrs. Sandhya Das, loving Children Monima and Prasenjit and nephew Subhajit for their encouragement and inspiration during the preparation of the manuscript of the book. Sriman Prasenjit and Subhajit help me in various ways and full credit goes to them for editing and total makeup of typed copy in the computer and doing some photographs for the book.

The book will be in help to the students of Agriculture, Teachers, Agricultural Officers engaged in Extension works, and the persons engaged in Agricultural professions.

All Suggestions and criticism for further improvement of the book will be thankfully acknowledged. I shall consider my efforts worthy and successful if this book serves the purposes for which it is written.

Lastly, I express my gratitude to Mr. Tanay Sharma, Managing Director, Scientific Publishers (India), Jodhpur and his staff engaged for the publication of the book so efficiently and promptly.

P. C. Das

Arya Pally (North),
Baranilpur,
P.O. SRIPALLY,
Dist. BURDWAN 713103 (W.B.)
Mobile : 9434349483
E-mail : pcdas46@gmail.com

CONTENTS

INTRODUCTION

Spices are very important as food and as medicine, and they bring out the unique natural taste in cuisines and can be used to change the look of the food to make it more attractive. Sometimes, spice is used to hide other flavours. Spices are essentially the flavoured or aromatic part of a plant used as a preservative or for food flavouring. Indian spices include a variety of spices that are grown across the Indian subcontinent (South Asia). In the Indian subcontinent, spice is a dried seed, fruit, root, bark or vegetative substance and it is primarily used for flavouring, colouring or preserving food. Spice is a broad term used to describe herbal by–products that add flavour, and aesthetic, aromatic and therapeutic treatment to food, drink and other items. Spices are spread over to most of the tropical part of the world. With different climates in different parts of the country, India produces a variety of spices, many of which are native to the Subcontinent, while others were imported from similar climates and have since been cultivated locally for centuries. Spices play an important role in the nutrition of our daily diet. Scientists have done a lot of research on this and have found out that spices contain more antioxidants than fruits and vegetables. Spices contain more antioxidants when they are dried than when they are raw. Plants that are the source of spice became important cash crops over the centuries and since their introduction, their uses have been multiplied. Spices are the major commercial crops in India and earn a major part of foreign exchange annually. The spices have been the backbone of agricultural industry.

India is the native home of many important spices such as Black pepper, cardamom, tamarind, curry leaf and to some extent Ginger, turmeric, garcinia, and Cinnamon, where a good variability exists. In fact, there is no State in India that does not grow spices, which in turn, plays an important role for the lives of the people and for their economic stability.

Spices are used for the production of different products as follows:

- Cosmetic
- Perfumes
- Pharmaceuticals

- Hygienic products

1.1. USES OF SPICES

- The principle use of spices is to season the foods to impart flavour, aroma and taste.
- Spices are also used as preservatives and fumigants.
- Spices also find use in pharmacy and indigenous medicines.
- Spices are also used in perfumery, soaps, cosmetics, tooth paste, confectionery, incenses, dyes etc.

1.2. CLASSIFICATION OF SPICES

There are about 70 spices grown all over the world, which are classified in different ways.

1.2.1. Botanical classification (family wise)

- Apiaceae - Anise, Coriander, Fennel, Cumin, Celery, Asafoetida.
- Cluciaceae - Garcinia.
- Fabaceae - Fenugreek.
- Lauraceae - Bay leaf, Cinnamon.
- Liliaceae - Garlic, Onion.
- Myrtaceae - Allspice, Clove, Nutmeg.
- Piperaceae - Black pepper.
- Rutaceae - Curry leaf.
- Zingiberaceae - Ginger, Cardamom, Turmeric.

1.2.2. Depending on the plant part useful as spice

- Aril spices - Mace.
- Bark spice - Cinnamon, Cassia.
- Bulbous spices - Onion, Garlic.
- Fruit spices
  - Berries - Pepper, Allspice.
  - Capsules - Cardamom, Chilli.
- Leafy spices - Mint, Coriander, Methi.
- Rhizome spice - Turmeric, Ginger.
- Root spice - Angelica, Horseradish.

- Seed spices - Coriander, Celery, Methi, Mustard, Cardamom, Cumin, Fennel, Fenugreek, Dill, Caraway and Anise

1.2.3. Depending upon the longevity of spice plants

- Annual spice - Coriander, Mint, Methi.
- Biennial spices - Onion, Garlic.
- Perennial spices - Clove, Nutmeg, Pepper.

1.2.4. Depending upon the type of the plant

- Bush spices - Cardamom.
- Climber spices - Pepper, Vanilla.
- Herbaceous spices - Coriander, Fenugreek, Fennel, Cumin.
- Herbal spices - Saffron, Lavender, Thyme, Oregano, Celery, Anise, Sage, Basil.
- Tree spices - Clove, Nutmeg, Cinnamon, Cassia.

1.2.5. Depending upon the economic importance i.e., magnitude of trade, earnings and use

- Major spice - Pepper, Cardamom, Turmeric, Ginger, Clove, Nutmeg.
- Minor spice - Coriander, Fenugreek, Cumin, Fennel, Capsicum, Onion, Garlic, Tamarind, Vanilla.

The different parts of spice crops used as spices are as follows:

- Dried fruits - Black pepper, Cardamom, Cumin, Fennel, Celery.
- Seeds - Mustard, Fenugreek, Cumin, Cardamom, Coriander, Anise, Dill, Ajowan.
- Dried flowers - Cloves, Saffron, Asafoetida.
- Stigma - Saffron.
- Bulb - Onion, Garlic, Fennel.
- Bark - Cassia, Cinnamon, Bay leaf.
- Roots and rhizome - Ginger, Turmeric.
- Leaves - Fenugreek, Fennel, Coriander, Basil, Mint, Parsley, Bay leaf.
- Buds - Cinnamon, Celery.

1.3. PROPERTIES OF SPICES

- Spices are well known as appetizers.

- Spices add a tang (taste) and flavor to otherwise insipid foods.
- Some of spices also possess antioxidant properties.
- Some of spices have preservative qualities (clove and mustard).
- Some spices have strong antimicrobial and antibiotic activities.
- Several of spices possess medicinal properties.

Source: Hort - 282. Acharya N.G. Ranga Agricultural University.

Table 1. List of spice crops and the parts used as spices

S.N.	English common name	Botanical name	Family	Part used as spices
1.	Cardamom (Small)	*Elettaria cardamom* Maton	Zingiberaceae	Fruit and seed
2.	Cardamom (Large)	*Amomum subulatum* Roxb.	Zingiberaceae	Fruit and seed
3.	Pepper	*Piper nigrum* L.	Piperaceae	Fruit
4.	Bird's Eye	*Capsicum frutescens* L.	Solanaceae	Fruit
5.	Capsicum	*Capsicum frutescens* L.	Solanaceae	Fruit
6.	Chilli	*Capsicum frutescens* L.	Solanaceae	Fruit
7.	Paprika	*Capsicum frutescens* L.	Solanaceae	Fruit
8.	Ginger	*Zingiber officinale* Rosc.	Zingiberaceae	Rhizome
9.	Turmeric	*Curcuma longa* L.	Zingiberaceae	Rhizome
10.	Coriander	*Coriandrum sativum* L.	Apiaceae	Leaf and fruit
11.	Cumin	*Cuminum cyminum* L.	Apiaceae	Fruit
12.	Fennel	*Foeniculum vulgare* Mill.	Apiaceae	Fruit
13.	Fenugreek	*Trigonellafoenum graecum* L.	Fabaceae	Seed
14.	Celery	*Apium graveolens* L.	Apiaceae	Leaf, fruit and stem
15.	Aniseed	*Pimpinella anisum* L.	Apiaceae	Fruit
16.	Ajowan	*Trachyspermum ammi* L.	Apiaceae	Fruit
17.	Caraway	*Carum carvi* L.	Apiaceae	Fruit
18.	Dill	*Anethum graveolens* L.	Apiaceae	Fruit
19.	Cinnamon	*Cinnamomum zeylanicum* Breyn	Lauraceae	Bark
20.	Cassia	*Cinnamomum cassia* Blume	Lauraceae	Bark
21.	Garlic	*Allium sativum* L.	Alliaceae	Bulb
22.	Curry leaf	*Murraya koenigii* (L) Sprengel	Rutaceae	Leaf
23.	Kokam	*Garcinia indica* Choisy	Clusiaceae	Rind

24.	Mint	*Mentha piperita* L.	Lamiaceae	Leaf
25.	Mustard	*Brassica juncea* L. Czern	Brassicaceae	Seed
26.	Parsley	*Petroselinum crispum* Mill.	Apiaceae	Leaf
27.	Pomegranate	*Punica granatum* L.	Punicaceae	Seed
28.	Saffron	*Crocus sativus* L.	Iridaceae	Stigma
29.	Vanilla	*Vanilla planifolia* Andr.	Orchidaceae	Pod
30.	Pepper Long	*Piper longum* L.	Piperaceae	Fruit
31.	Star Anise	*Illicium verum* Hook.	Illiciaceae	Fruit
32.	Sweet flag	*Acorus calamus* L.	Araceae	Rhizome
33.	Greater Galanga	*Alpinia galanga* Willd.	Zingiberaceae	Rhizome
34.	Horseradish	*Armoracia rusticana* Gaertn.	Brassicaceae	Root
35.	Caper	*Capparis spinosa* L.	Capparidaceae	Flower buds
36.	Clove	*Syzygium aromaticum* (L) Merr.& Perry	Myrtaceae	Unopened flower bud
37.	Asafoetida	*Ferula asafoetida* L	Apiaceae	Oleogum resin from rhizome and thickened root
38.	Camboge	*Garcinia cambogia* (Gaertn). Desr	Clusiaceae	Rind
39.	Hyssop	*Hyssopus officinalis* L.	Lamiaceae	Leaf
40.	Juniper berry	*Juniperus communis* L.	Cupressaceae	Berry
41.	Bay leaf	*Laurus nobilis* L.	Lauraceae	Leaf
42.	Lovage	*Levisticum officinale* Koth.	Apiaceae	Leaf and stem
43.	Marjoram	*Marjorana hortensis* Moench.	Lamiaceae	Leaf
44.	Nutmeg	*Myristica fragrans* Houtt.	Myristicaceae	Seed
45.	Mace	*Myristica fragrans* Houtt.	Myristicaceae	Aril
46.	Basil	*Ocimum basilicum* L.	Lamiaceae	Leaf
47.	Poppy seed	*Papaver somniferum* L.	Papaveraceae	Seed
48.	Allspice	*Pimenta dioica* (L) Merr.	Myrtaceae	Fruit and leaf
49.	Rosemary	*Rosmarinus officinalis* L.	Lamiaceae	Leaf
50.	Sage	*Salvia officinalis* L.	Lamiaceae	Leaf
51.	Savory	*Satureja hortensis* L.	Lamiaceae	Leaf
52.	Thyme	*Thymus vulgaris* L.	Lamiaceae	Leaf
53.	Oregano	*Origanum vulgare* L.	Lamiaceae	Leaf
54.	Tarragon	*Artemisia dracunculus* L.	Asteraceae	Leaf
55.	Tamarind	*Tamarindus indica* L.	Caesalpiniaceae	Fruit

Source: http://www.evergreenexports.net/spices-botanical-names.asp

Table 2. Medicinal properties of some spice crops

S.N.	Name of spice	Medicinal value
1.	Asafoetida (Hing)	It is used to treat whooping cough and stomach ache.
2.	Bay leaf (Tejpatta)	Bay leaf possesses antifungal and antibacterial property.
3.	Cardamom (Elaichi)	It is useful to control bad breath and digestive disorder. A whole Cardamom chewed is good for coping diabetes and nausea.
4.	Chilli (Lal mirch)	The antioxidants in chilli cope with cholesterol.
5.	Cinnamon (Dalchini)	It supports natural production of insulin and reduces blood cholesterol. It is used as an antidote for diarrhoea and stomach upset, and for blood sugar regulation. Taking half teaspoon of Cinnamon powder every day keeps cancer risk away. It is helpful in reducing tumour growth. Cinnamon, a natural food preservative, is a source of Iron (Fe) and Calcium (Ca). Cinnamon has antimicrobial properties and is a rich source of antioxidant. It prevents the growth of bacteria, fungi and yeasts. It is also effective in diabetes.
6.	Clove (Laung)	Clove oil is beneficial for coping with tooth ache and sore gum. It is also beneficial for chest pain, fever, digestive problem, and cough and cold.
7.	Coriander (Dhaniya)	Coriander can be used externally in aching joints and rheumatism. It is good for coping with sore throat, allergies, digestion problem, hay fever etc.
8.	Cumin (Jeera)	Cumin is a good source of iron and keeps the immune system healthy. It is antidiabetic, antioxidant and antimicrobial. Water boiled with Cumin seeds is good for coping with dysentery. Cumin seeds contain a compound called 'Thymoquinone' that checks the proliferation of cell responsible for prostate cancer.
9.	Curry leaf (Curry patta)	Green leaves are beneficial for reducing blood sugar. Dried leaves are extensively used in herbal medicine.
10.	Fenugreek (Methi)	It increases libido and male potency, and cures skin problem. Fenugreek seed is good for increasing breast milk and also helpful for treating diabetes and lowering cholesterol.
11.	Garlic (Lassan/Rasun)	Garlic is useful in cough and cold, abdominal pain, high blood pressure, diarrhea, diabetes, eye infection, whooping cough, catarach, menstruation problem, asthma, dandruff, abortion, tuberculosis, vomiting and possesses antibiotic property. Regular consumption of garlic decreases blood pressure, rheumatism, cholesterol and liver diseases.
12.	Ginger (Adrak)	It is used to treat digestive problem, cough and cold. It inhibits nausea or vomiting caused by morning sickness or motion sickness.

		Ginger is also beneficial for arthritis, cancer and Alzheimer's disease for its inflammatory properties.
13.	Mustard (Sarisa)	It is a good source of Iron (Fe), Zinc (Zn), Magnesium (Mg), Calcium (Ca) and protein. Mustard oil cures rheumatism.
14.	Nutmeg (Jaiphal)	It is helpful in the treatment of asthma, heart disease and bad breath.
15.	Onion (Piyaaz)	It is used to treat diabetes, hypertension, cough and cold, fever and abscess. Bulb is helpful for alleviate menstrual and uterine pain.
16.	Pepper (Kali mirch)	The Pepper helps to deal with cold, cough, infection, muscle pain and digestive problems.
17.	Saffron (Zaffran/Kesar)	It is good remedy for cough and cold, asthma and skin diseases. Saffron contains a natural caratenoid dicarboxil acid called 'Crocetin', which is the primary cancer–fighting element.
18.	Star Anise (Chakra phool)	Oil is beneficial for rheumatism and it is helpful for digestion and avoiding bad breath.
19.	Turmeric (Haldi)	Turmeric powder can be used for healing cuts and wounds. It also helps coping with diabetes. It contains powerful polyphenol curcumin that has been clinically proven to retard the growth of cancer cells causing prostate cancer, melanoma, breast cancer, brain tumour, pancreatic cancer and leukaemia among a host of other. The curcumin promotes 'Apoptosis' (programmed cell death/cell suicide) that safely eliminates cancer breeding cells without posing a threat to the development of other healthy cells. Turmeric has anti–inflammatory properties and is helpful in Crohn's disease, ulcerative colitis, cystic fibrosis, cancer, rheumatoid arthritis, cardiovascular and Alzheimer's disease. Interestingly, Turmeric is traditionally used in all Indian preparations to add bright yellow colour to the food.

1. CARDAMOM

Scientific name - *Elettaria cardamomum*

Family - Zingiberaceae

Commercial part - Seed

Indian names - *Bengali* : Chhoti elachi; *Hindi* : Chhoti elaichi; *Gujarati* : Elaychi; *Kannada* : Yelakki; *Kashmiri* : Aalbuduaal; *Malayalam* : Elathari; *Marathi* : Velchi; *Oriya* : Alaichi; *Punjabi* : Elaychi; *Sanskrit* : Ela; *Tamil* : Yelakkai or Elakkai; *Telugu* : Yealak-Kayulu or Elakkayi; *Urdu* : Ilaychi

Cardamom (*Elettaria cardamomum* L.), popularly known as '*Queen of Spices*', is native to the evergreen rainy forests of Western Ghats in South India. It is cultivated in Kerala, Karnataka and Tamil Nadu. Cardamom is used for flavouring various preparations of food, confectionery, beverages and liquors.

Cardamom is a plant species and a spice is made from the seeds of the plant. The word 'Cardamom' is derived from the Latin *Cardamomum,* itself the latinization of the Greek *kardamomon,* a compound of *kardamon,* 'cress ov*amomon',* which was the name for a kind of an Indian spice plant. The useful part of Cardamom is the dried mature fruit. It is usually referred to as the capsule. Cardamom is used for flavouring curries, cakes, and bread and for other culinary purposes. It is also used as masticatory and for flavouring coffee and confectionery. India is the largest producer and exporter of this spice, accounting for more than 60 per cent of the total world production and total world trade. The demand for large Cardamom in the export market is increasing steadily. Cardamom is often called 'the third most expensive spice in the world' (after saffron and vanilla) and the high price reflects the reputation of this most pleasantly–scented spice. Cardamoms are members of the Ginger family, the Zingiberaceae and both - the entire seed pods and the seeds themselves, can be used as a spice.

Cultivation of Cardamom is mostly concentrated in the evergreen forests of Western Ghats in South India. Besides India, Cardamom is grown as a commercial crop in Guatemala and on small scale in Tanzania, Sri Lanka, El Salvador, Vietnam, Laos, Thailand, Cambodia, Honduras, and Papua and New

Guinea. Green cardamom is a spice that is derived from the Ginger family of herbs and is known for its spicy–sweet taste and strong aroma. The pods of the Cardamom plant, which are also used in Black cardamom and Madagascar cardamom, must be dried and stored carefully because they rapidly lose their taste and scent after being ground. Green cardamom is slightly more expensive than most spices.

1.1. COMPOSITION OF CARDAMOM

The Cardamom contains different food ingredients as follows:

- Carbohydrates
- Proteins
- Crude fibre
- Vitamins such as vitamin C, niacin, pyridoxine, riboflavin and thiamin
- Minerals such as phosphorus, copper, iron, calcium, magnesium, manganese and zinc
- Essential oil (volatile oil) such as sabinene, cineol, limonene and terpineol in the form of formic and acetic acids

1.2. ORIGIN AND DISTRIBUTION

India is considered as the native home of *Elettaria cardamomum,* though the major centre of diversity for the genus is the Sarawak (Malaysia) and Borneo region, from where eight species have been listed (Sakai and Nagamasi 2001). Natural population of Cardamom now exists only in the evergreen forests of Western Ghats. In India, genus *Elettaria* is represented only by one species *(Elettaria cardamomum).* Cardamom cultivation in India is confined to three States; prominent among them is Kerala while Karnataka and Tamil Nadu are the other two Cardamom–growing States.

1.3. BOTANICAL DESCRIPTION

- **Plant** - The plant is a herbaceous perennial (2–5 m in height) with underground (subterranean) rhizomes and the aerial shoot is a pseudostem and formed by the encircling leaf sheaths.
- **Stem** - The tuberous underground rhizome is its real stem.
- **Leaf** - The leaves are long, alternate and lanceolate in shape.
- **Flower** - Flowers are arranged in clusters (known as *cincinni*) subtended by scale leaves. Flowers are bisexual, bracts linear, oblong and persistent. Flowers are borne on panicles and they emerge directly from the underground stem on long floral stalks.

- **Inflorescence** - Inflorescence is a long panicle arising from the underground stem, but comes up above the soil. The linear growth of panicles extends over a period of about seven months. The growth habit of the panicles and the shape as well as the size of the capsules varies in different cultivated varieties/types of Cardamom. Flowers are arranged in clusters (known as *cincinni*) subtended by scale leaves. Normally, flowering in Cardamom is seen throughout the year on panicles produced during the current as well as in the previous year.

Fig. 1. Grown–up Cardamom plant

Cardamom is a perennial, herbaceous, rhizomatous plant. The following is some information about Cardamom:

- **Season** - April–September.
- **Flowering and fruiting** - Four months from April to May and continues up to August–September. Peak flowering will be in the months of May–June.
- **Bearing period** - Cardamom plants start bearing in 2–3 yr after planting or maximum at 6–10 yr and steady yield throughout.
- **Harvesting** - September–January; peak period - late October to mid December. From flowering to maturity, the fruit takes about 5–6 months.
- **An average life of a plant** - 20 years.

1.4. CLIMATE

The main conditions for growing large Cardamom are as follows:

- **Temperature range** - Maximum - 14°C–33°C; Minimum - 4°C–22°C
- **Annual precipitation** - 200–250 cm, well distributed throughout the year. Cardamom thrives best in evergreen tropical forest at an altitude of 765–1,675 m above MSL. Cool, constantly moist but not waterlogged situations and an eastern or south–eastern aspect are considered favourable for the crop.

Fig 2. (1) Cardamom plant (2) Fruit

1.5. SOIL

Forest loamy soils are preferred for Cardamom cultivation. The soil that is generally acidic in nature (Soil pH 4.2–6.8) is good for the cultivation of Cardamom. The crop grows well on well–drained, fertile soil, forest loams and red, deep, good–textured lateritic soils having plenty of humus.

Soil nutrient status - High in organic matter and nitrogen, low to medium in available phosphorus and medium to high in available potassium.

1.6. LAND PREPARATION

The soil should be brought to fine tilth by ploughing the land by tractor or powertiller. The tillage must be followed by planking to level the soil and to break the clods. Compost or Farm yard manure (FYM) should be applied during land preparation and well mixed by ploughing followed by planking. The land should have drainage facility as this crop does not grow in waterlogged soil. The pits, measuring 45 x 45 x 45 cm should be dug in the prepared field. The pits should be partly filled with a mixture of rich surface soil and compost or FYM.

1.7. PROPAGATION

Cardamom can be propagated vegetatively by sucker and by seedlings.

- **Vegetative propagation**

For vegetative propagation, a section of the rhizome called 'bulb' with not less than three shoots and obtained from healthy 1.5–2.0 yr old and regularly high–yielding mother clumps, should be used for planting. The bulb should be planted at a spacing of 0.6 m in the trenches during the month of March–October. Plants propagated vegetatively come to bearing one year earlier than the seedling propagated plant.

- **Propagation through seed**

Ripe capsules of the desired cultivar should be collected from high–yielding plants during September–October. The seeds should be extracted by gently pressing the capsules. The seeds should be sown in the nursery bed for raising seedlings immediately after extraction in order to ensure uniform and early germination. This nursery bed should be 4–6 m long, 6 m wide and 25–30 cm high. The seed bed should receive adequate amount of compost or FYM. The seeds may be sown in the nursery bed by broadcasting or in lines and the sown seed should be lightly covered with fine soil and leaf mulch. Nowadays, the seedlings are also raised in polythene bags containing rich forest soil.

1.8. PLANTING

The seedlings, when about three months old, should be transplanted in the second nursery for at least one year. The seedlings should be provided adequate water and shade by thatching the nursery bed. Then, one or two seedlings should be planted in the main field in the month of June–July or alternatively in the month of September–October. The seedlings should be sown 2–3 inches deep and the soil around the seedlings should be well pressed. The seedlings need to be securely tied to a stout wooden stake to prevent them from falling down by strong monsoon rains and winds.

Spacing : Row to row - 2 m

Plant to plant - 2 m

The mixed cropping of Cardamom with the plantation of arecanut, rubber and tea is gaining popularity. The Cardamom gets benefit from the cultural and manurial treatments that are given to the plantation crops.

1.9. MANURING

Manures

Compost or FYM - 10.0–15.0 t/ha

Fertilizers

Table 3. Fertilizers recommendation for Cardamom

S.N.	Condition	Amount of nutrients (kg/ha)		
		Nitrogen (N)	Phosphorus (P_2O_5)	Potassium (K_2O)
1.	Rainfed condition	30.0	30.0	60.0
2.	Irrigated condition	75.0	75.0	150.0

Method of application

The fertilizer should be applied in split doses as follows:

- **First application** - The first dose of fertilizers should be applied during the month of May, which helps in the production of suckers and development of capsules.
- **Second application** - The second dose of fertilizers should be applied during late September, which helps the initiation of panicles and suckers. Only half the dose of fertilizer is to be applied during the first year and full dose should be given from second year onwards.

Cardamom being a surface feeder, deep placement of fertilizers is not advisable. The fertilizer is applied at a radius of 30 cm and covered with a thin layer of soil.

Table 4. Released varieties of Cardamom with yield and quality characteristics

S. N.	Variety	Source	Yield (kg/ha)	Essential oil %	1,8 cineole %	α -Terpenyl acetate %	Capsule shape	Areas recommended for cultivation
1.	IISR Coorg Suvasini	IISR, CRC Appangala	409	8.7	42	37	Oblong	Kodagu and Hassan districts of Karnataka
2.	PV-1	KAU, Pampadumpara	260	6.8	33	46	Long	All Cardamom–growing tracts of Kerala and Karnataka
3.	Mudigree-1	UAS, Bangalore	275	8.0	36	42	Oval	Malnad region of Karnataka
4.	Mudigree-2	UAS, Bangalore	476	8.0	45	38	Round	Traditional Cardamom–growing tracts of hill zones of Karnataka
5.	ICRI-1	ICRI, Myladumpara	325	8.3	29	38	Round	South Idukki zone of Kerala

6.	ICRI-2	ICRI, Myladumpara	375	9.0	29	36	Oblong	Vandanmettu and Nelliam-pathi zones of Kerala
7.	ICRI-3	ICRI, Myladumpara	439	6.6	54	24	Oblong	Hill zones of Karnataka
8.	ICRI-4	ICRI, Thadiyankudisai	455	6.4	-	-	Globose	Lower Pulneys in Tamil Nadu
9.	IISR Avinash	IISR, CRC, Appangala	847	6.7	30.4	34.6	Oblong	Rhizome rot–infested areas
10.	IISR Vijetha -1	IISR, CRC, Appangala	643	7.9	44.9	23.4	Oblong	Moderate to high shaded mosaic–infes-ted areas
11.	PV-2	KAU, Pampadumpara	982	10.45	-	-	Long	Cardamom hill reserves of Kerala
12.	Njallani Green Gold	Farmers' Selection	1600	-	-	-	bold	All Cardamom–growing regions

Source: Advances of Spices Research, Agrobios, Jodhpur

Table 5. Promising Cardamom hybridization derived lines evolved at ICRI, Myladumpara

Hybrid combinations	Projected yield (kg/ha)
MCC 16 x MCC 40	610
MCC 61 x MCC 40	675
MCC 21 x MCC 16	650
MCC 21 x MCC 40	870
MCC 16 x MCC 61	800

Table 6. High–yielding varieties and selections in Cardamom

S.N.	Selection/ Variety	Cultivar	Special distinguishing characteristics	Area of adaptability	Source
1.	ICRI-1	Malabar	An early–maturing, profu-sely–flowering variety, medium size panicle with globose, extra bold dark-green coloured capsule.	South Iduki Zone	ICRI (Spices Board), Myladumpara, Kerala - 695553
2.	ICRI-2	Mysore	Performs well under irrigated condition, suitable for higher altitude, medium long	Vadammedus and Nelliampathy of Kerala and	ICRI (Spices Board), Myladumpara, Kerala - 695553

			panicle, oblong, bold and parrot green capsule, tolerant azhukal disease	Annamalai and Meghamalai of Tamil Nadu	
3.	ICRI-3	Malabar	An early–maturing type, non–pubescent leaves, oblong bold, parrot green capsule, tolerant to rhizome rot disease	Cardamom growing tract of Karnataka	Regional station, ICRI (Spice Board), Sakles-pur kandraya - 573134
4.	TDK-4		An early– maturing variety, adaptable to low rainfall area, medium size panicle, globose, bold parrot green capsule, non–pubescent leaves	Adapted to lower Pulney hills of Tamil Nadu, suitable for low rainfall area (1,500 mm) and having similar agroecological conditions	Regional station, ICRI (Spice Board), Thadian kudisai -62421, Tamil Nadu.
5.	TDK-11	Malabar	An early– maturing variety, long panicle, oblong bold and parrot green capsule	Adapted to less rainfall conditions of lower Pulney hills of Tamil Nadu	Regional station, ICRI (Spice Board), Thadian kudisai-62421, Tamil Nadu.
6.	PV-1	Malabar	An early–maturing variety with slightly–ribbed light green capsule, short panicle, close raceme, narrowly ellipsoid to elongated capsule	All Cardamom–growing tracts in Kerala and Parts of Tamil Nadu	Cardamom Research Station, Kerala Agricultural University, Panpadumpara, Kerala - 685566
7.	CCS-1	Malabar	An early– maturing variety, suitable for high density planting, long panicle, oblong bold parrot green–coloured capsule	All Cardamom growing tract of Karnataka and Wynal of Kerala	Indian Institute of Spice Rese-arch, Regional Centre (ICAR), Appangala, Kar. - 577132
8.	Mudigere-1	Malabar	Compact plant, tolerant to hairy caterpillar and white grubs, short panicle, bold, pale green capsule, tolerant to thrips and shoot borer, pubescent leaves	In the traditional Cardamom–growing Malanad areas of Karnataka	Regional Rese-arch Station, University of Agricultural Science (UAS), Mudigere, Karnataka - 577132

9.	MCC-200 (Nijallani)	Vazhukka	Non–pubescent, semi–erect, globose extrabold and dark green capsules	Suitable to Kerala State	Private nurseries in Idukki district of Kerala
10.	MCC-12	Vazhukka	Semi–erect panicles and dark, deep green capsules suitable under sparse shade condition	Oblong capsules suitable under sparse shade condition of Kerala	ICRI, Spice Board, Myladumapara, Idukki - 685553
11.	MCC-16	Vazhukka	Profusely branched, compound and semi–erect panicles, globose green capsules, recom-mended for irrigated and swampy areas	Adapted to Kadamkuzhy and Umdum-banchola zones of Kerala	ICRI, Spice Board, Myladumapara, Idukki - 685553
12.	MCC-40	Malabar	An early–bearing variety, globose bold and green capsule	Suitable for all Cardamom–growing tract of Kerala	
13.	BR-1	Malabar	Resistant to rhizome rot disease, suited for hot spots of rhizome rot and leaf blight disease, high quality elongated capsule	Kodagu, North Wayanad, Hassan and Chikmagarbir	IISRl Research Centre, Appangalakarn
14.	Mudigere-2	Malabar	Suitable for cultivation in valleys in Karnataka	Cardamom–growing tract of Karnataka	Regional Rese-arch Station, University of Agricultural Science, Mudigere

Source: Spice Board of India, Ministry of Commerce and Industry, Government of India

1.10. INTERCULTURAL OPERATION

1. Weeding

In the first year of planting, frequent weeding is necessary to eliminate the competition between young Cardamom plants and the weeds. The land should be kept free from weeds as they pose a problem in the cultivation of Cardamom. Weeding needs to be done twice or thrice depending on the intensity of weeds to keep land free from weeds. The weeding should be done as per the following schedule:

- **First weeding** - The first weeding should be carried out in the month of May–June.

- **Second weeding** - The second weeding should be carried out in the month of August–September.
- **Third weeding** - The third weeding should be carried out in the month of December–January. Thereafter, depending upon the weed intensity, 1–3 weedings per annum would be necessary. The weeds removed can be utilized as mulch around the plant.

Weeds can also be controlled by spraying Glyphosate 41 SL @ 625 ml in 500 l of water per hectare in the interspaces between rows leaving 60 cm around the plant base.

2. Mulching

Mulching is an important cultural operation in Cardamom. Fallen leaves of the shade trees or uprooted weeds are utilized for mulching. Mulches should be applied during November–December to reduce the ill–effects of drought conditions during ensuing summer.

3. Trashing (Cleaning)

Trashing consists of removal of old and drying shoots of the plant once in a year with the onset of monsoon rains under rained conditions and 2–3 times under irrigated conditions. Weeding and clearing may be done simultaneously during May–June to promote the healthy growth of new shoots. A second and final clearing may be done during August–September to prevent the damage to fruits by rodents.

4. Racking and digging

A light raking or digging of soil should be given around the plant up to a radius of 60–75 cm, to conserve the moisture for the ensuing dry period particularly in low rainfall areas. This should be done every year after the rainy season.

5. Earthing up

Earthing up is found beneficial to encourage new growth. For this, a thin layer of fertile soil, rich in organic matter, may be earthed up at the base of the plant, up to collar region.

6. Shade regulation

Cardamom is ecologically classified as 'Pseophytes' i.e., shade–loving plant. Cardamom thrives best under moderate–natural shade. The growth, flowering and fruiting are adversely affected under exposed condition. The shade tree such as *Erythrina* species shouldbe planted for providing temporary shade in the earlier years. Palangi (*Artocrdamon fraxinifolius*), Red cedar (*Ceorelle tonnav*), Karimaram (*Diospyros albenum*) and Elangi (*Mimusops elangi*) are common shade trees in Cardamom estates. Shade regulates soil moisture as well as temperature, and provides congenial microclimate for Cardamom. Shade protects the Cardamom plants from sun–scorching, rains and winds.

1.11. IRRIGATION

Irrigation should be given as per requirement of the plants. The interval of irrigation depends on the water–holding capacity of soil and the weather condition. In order to overcome the dry spell during summer, it is necessary to irrigate the crop to get maximum production as it helps in initiation of panicles, flowering and fruit set. They may be irrigated at an interval of 10–15 days till the onset of monsoon. Sprinkler and drip irrigation is found beneficial for this crop.

1.12. HARVESTING

Cardamom plants normally start bearing two years after planting and reach the full bearing stage at the age of seven or eight years. In most of the areas, the peak period of harvesting is during October–November. Ripe capsules should be harvested in order to get maximum green colour during curing. Picking should be carried out at an interval of 15–25 days. Yields decline from ten years.

1.13. PROCESSING

The commercial product of Cardamom is the dried capsules. At the time of harvesting, the capsules are juicy and fleshy, so they must be cured before sending them to the market. The harvested capsules should be dried either in fuel kiln or electrical drier or in the Sun. The capsules kept for drying should be spread thinly and stirred frequently to ensure uniform drying. The dried capsules should be rubbed with hands or coir mat or wire mesh and winnowed to remove any foreign matter. If drier is used, it should be dried at 45–50°C for 14–18 hours, while for kiln; overnight drying at 50–60°C is required.

1.14. BLEACHING

Green colour of the Cardamom capsule plays a vital role in the market. Green colour of the capsules can be preserved by alkali treatment. It has been found that soaking the freshly–harvested green Cardamom capsules in 2% washing soda (Na_2CO_3) solution for 10 min prior to drying, helps in retaining the green colour during drying.

1.15. STORAGE

The capsules are sorted out according to size and colour, and then stored in black polythene–lined gunny bags to retain the green colour during storage. The bags containing the capsules of Cardamom should be kept in well–ventilated godowns.

1.16. YIELD

50.0–70.0 kg/ha (average). The yield varies with age and variety of the Cardamom as follows:

Table 7. Yield of Cardamom

Sr. No.	Year of bearing	Yield (Dry capsules) (Kg/ha)
1.	First year	25.0–50.0
2.	Second year	50.0–70.0
3.	Fifteen years	70.0–100.0

1.17. USES

The pods of Cardamom can be used in whole or split when cooked. There are various uses of Cardamom as follows:

- Due to its pleasant aromatic odour, large Cardamom is used for flavouring various vegetables and meat preparation in Indian dishes. It is also used as a flavouring agent in confectionery, hot or sweet pickles, curries, rice pudding and other rice dishes, and in beverages.
- Cardamom is stimulant, carminative and digestive. Cardamom is used mainly to treat colds, bronchitis, fevers, inflammatory conditions and liver complaints; however, there is no evidence to support any of these uses.
- The seeds can be chewed for good flavour in the mouth and to get rid of bad breath.

1.18. PLANT PROTECTION

Insect pests

1. Thrips

Scientific name - *Sciothrips cardamom*

The thrips is the most destructive and persistent insect pest of Cardamom. Adults and nymphs of the insect cause damage to panicles, flowers and capsules.

Nature of damage

The adults and nymphs suck the sap and causes damage to leaves, shoots, inflorescence and fruits. As a result, the panicles become stunted and premature shedding of flowers and immature capsules occurs, which causes reduction in formation of capsules. Infestation causes formation of corky encrustations on pods resulting in their malformed and shrivelled condition. Affected capsules appear malformed, shrivelled and sometimes with gaping slits.

Control measures

- The dry drooping leaves, dry leaf sheaths, old panicles and other dry plant parts should be destroyed by burning.
- The thrips can be controlled by spraying any of the insecticides in the Table 8.

Table 8. Insecticides used for controlling the Thrips of Cardamom

Sr. No.	Name of insecticide	Amount (per litre of water)
1.	Acephate 75 WP	0.75 g
2.	Quinalphos 25 EC	2.0 ml
3.	Clothianidin 50 WDG	0.1 gm
4.	Triazophos 40 EC	2.0 ml
5.	Neem oil 3000 ppm	3.0 ml

2. Shoot, panicle and capsule borer

Scientific name - *Dichocrocis (Conogethes) punctiferalis*

Infestation of borers on shoot, panicle and capsule is a serious problem to Cardamom growers of Kerala, Tamil Nadu and Karnataka. Pest infestation is pronounced in three seasons – January–February, June-July, and September–October.

Nature of damage

The larvae at early stage bore the unopened leaf bud and panicle leading to drying up of the panicle from the affected spot and feed on young seed inside rendering the capsule empty. The larvae at late stage bore the pseudostem and feed on the central core of the stem, which results in the drying of terminal leaf and thus, produces characteristic 'dead heart'.

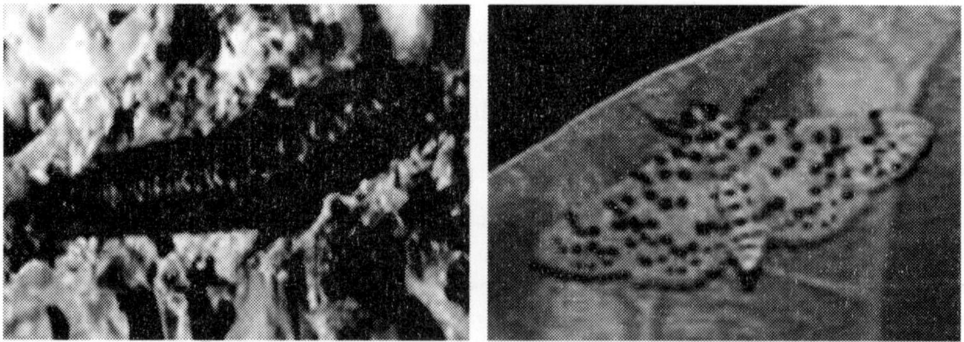

Larva Adult

Fig 3. Shoot, panicle and capsule borer

Control measure

- The adult moths rest on the lower surface of the Cardamom during day time. The affected Cardamom may be collected and destroyed by burning.
- The tillers showing 'dead heart' symptoms should be removed with due care and destroyed by burning.
- The insects can be controlled by spraying anyone of the insecticides in the Table 9.

Table 9. Insecticides used for controlling the shoot, panicle and capsule borer of Cardamom

Sr. No.	Name of insecticide	Amount (per litre of water)
1.	Acephate 75 WP	0.75 g
2.	Thiomethoxam 25 WG	0.2 g
3.	Carbosulphan 25 EC	2.0 ml
4.	Imidachloropid 17.8 SL	0.2 ml
5.	Acetamiprid 20 SL	0.1 g

3. Capsule borers

Scientific name - Jamides sp

Nature of damage

The caterpillars bore and feed on the inflorescence, flower buds, flowers and capsules. Affected capsules become empty with a big circular hole, turn yellowish–brown, which decay and drop off in the rainy season.

Control measures

- The insect can be controlled biologically by application of NPV (Nuclear Polyhydrosis Virus).
- The insects can be controlled by spraying any one of the insecticides as given in Table 10.

Table 10. Insecticides used for controlling the Capsule borer of Cardamom

Sr. No.	Name of insecticide	Amount (per litre of water)
1.	Acephate 75 WP	0.75 g
2.	Thiomethoxam 25 WG	0.2 g
3.	Carbosulphan 25 EC	2.0 ml
4.	Imidachloropid 17.8 SL	0.2 ml
5.	Acetamiprid 20 SL	0.1 g

4. Beetle borer

Scientific name - *Onthophagus* sp

Nature of damage

Larva makes the bore hole, which is neat, circular, tiny and at right angles to the pericarp, mostly on the ventral side. When attack is on young immature capsules, they put forth matured appearance, decay and drop off. Adult beetle bores the flowers and feeds on tissues.

Control measures

The insect can be controlled by spraying any one of the insecticides in the Table 11.

Table 11. Insecticides used for controlling the Beetle borer of Cardamom

Sr. No.	Name of insecticide	Amount (per litre of water)
1.	Acephate 75 WP	0.75 g
2.	Thiomethoxam 25 WG	0.2 g
3.	Carbosulphan 25 EC	2.0 ml
4.	Imidachloropid 17.8 SL	0.2 ml
5.	Acetamiprid 20 SL	0.1 g

5. Root grub

Scientific name - *Basilepta fulvicorne*

Damage is caused by the grubs of the insect. Beetles are small, greenish–blue or blue or green, and are found in large numbers during May–July and October–February.

Nature of damage

Grubs feed on Cardamom roots in patches. This reduces the uptake of nutrients and leads to yellowing of leaves.

Control measures

Early stages of the grub, which are usually present in soil during May–June and September–October can be controlled by applying Phorate 10 G or Thimate 10 G @ 30–40 gm per clump.

6. Hairy caterpillars

Scientific name - *Eupterote cardamomi, E. canarica, E. fabia*

Nature of damage

Caterpillars are gregarious in habit and they congregate on the trunks of shaded trees during day time and drop down on the Cardamom plants during night time, and feed voraciously on leaves, and defoliate the whole Cardamom plant. The hairy caterpillar appears and causes damage to the plant during the months of October–December.

Control measures

The insect can be controlled by spraying any one of the insecticides given in Table 12.

Table 12. Insecticides used for controlling the Root grub of Cardamom

Sr. No.	Name of insecticide	Amount (per litre of water)
1.	Chloropyriphos 20 EC	2.5 ml
2.	Malathion 50 EC	2.0 ml
3.	Acephate 75 WP	0.75 g
4.	Carbaryl 50 WDP	2.5 g

7. White fly

Scientific name - *Dialeurodes cardamom*

Nature of damage

Nymphs and adults remain in colonies on lower leaf surface and suck the sap. The affected leaves become yellow in colour and dry off. Sooty mold develops due to honeydew secretion.

Control measures

- Deep fallow ploughing helps in the destruction of maggot.
- The insect can be controlled by spraying any one of the insecticides in the Table 13.

Table 13. Insecticides used for controlling the White fly of Cardamom

Sr. No.	Name of insecticide	Amount (per litre of water)
1.	Chloropyriphos 20 EC	2.5 ml
2.	Malathion 50 EC	2.0 ml
3.	Acephate 75 WP	0.75 g
4.	Carbaryl 50 WDP	2.5 g

- The adult whitefly can be controlled by keeping poisonous water in small earthen pots in different places of the field. This poisonous water is prepared with different ingredients as follows:

  Malathion 25 EC - 3 ml

  Molasses - 200 g

  Water - 2 l

8. Root grub

Scientific name - *Basilepta fulvicorne, Holotrichia serrata*

Among the different species, B. fulvicorne is found to cause more damage in Cardamom tracts.

Nature of damage

The grubs feed on the roots in the form of irregular scraping. In advanced stages, entire root system is found damaged resulting in drying and rotting depending on the season of attack. In the severely–infested plants, leaves turn yellow and dry. As a result, plant growth gets retarded.

Control measures

- The adult beetles should be collected by using hand nets during peak period of emergence i.e., during April–May and should be killed by burning or burrowing them into the soil.
- Application of Phorate 10G @ 20–40 g per clump is found effective in controlling this insect. Light raking of soil before the insecticide application is essential for effective control of root grub.

9. Cardamom aphid

Scientific name - *Pentalonia nigronervosa f. Caladii*

Nature of damage

Colonies of aphids are seen under concealed conditions inside leaf sheaths of the older pseudostems. Nymphs and adults suck up plant sap. The aphids act as a vector of the mosaic or 'Katte' virus of Cardamom.

Control measures

The insect pest can be controlled by spraying any of the insecticides given in Table14.

Table 14. Insecticides for controlling the Cardamom aphide

Sr. No.	Name of insecticide	Amount (per litre of water)
1.	Chloropyriphos 25 EC	2.5 ml
2.	Carbosulfan 25 EC	2.0 ml
3.	Acephate 75 WP	0.75 g
4.	Imidachloroprid b17.8 SL	0.2 ml

Diseases

1. Capsul rot (Azhukal)

C.O.: *Phytopthora parasitica* var. *nicotianae* / *Phytophthora palmivora*

Capsul rot, popularly known as 'Azhukal' in local language (Malayalam) meaning rotting, is perhaps the most serious fungal disease of Cardamom. Menon et al (1972) reported this disease for the first time from the plantation of Idukki district in Kerala.

Symptoms

Initially, rotting symptoms are observed on the fruit/capsule only and accordingly, it is named capsule rot.

Small water–soaked lesions appear on capsules, which expend imparting dull green to brownish colour capsule rots emitting foul smell. Dark brown lesions appear on rachis also. Water–soaked lesions appear on leaves, which expend. Occasionally, pseudo stems / roots also show rotting.

Control measures

- The affected capsules / spikes / clumps should be removed from the plantation before the onset of monsoon.
- The shade should be regulated by pruning branches of overhead shade trees.
- The disease can be controlled by spraying on foliage and spikes with any one of the fungicides given in Table 15, once during May–June and again during August–September.

Table 15. Fungicides used for controlling the capsule rot of Cardamom

Sr. No.	Name of fungicide	Amount (per litre of water)
1.	Copper oxychloride 50 WP	5.0 g
2.	Carbendazim 50 WP	1.0 g
3.	Mancozeb 75 WP	2.5 g
4.	Chlorothalonil 75 WP	2.0 g

2. Clump rot and damping off

C.O.: *Pythium vexans, Rhizoctonia solani*

Symptoms

Young seedlings in the nursery rot at the collar and die

Control measures

- The bio-control inoculum (*T. harzianum*) @ 50 g/m^3 should be applied in the nursery bed before sowing of the seeds.
- The disease can be controlled by spraying on foliage and spikes with any one of the fungicides given in Table 16, once during May–June and again during August–September.

Table 16. Fungicides used for controlling the clump rot and damping off of Cardamom

Sr. No.	Name of fungicide	Amount (per litre of water)
1.	Coper oxychloride 50 WP	5.0 g
2.	Carbendazim 50 WP	1.0 g
3.	Mancozeb 75 WP	2.5 g
4.	Chlorothalonil 75 WP	2.0 g

2 BLACK PEPPER

Scientific name - *Piper nigrum*

Family - Piperaceae

Commercial part - Fruit

Indian names - *Bengali* : Gol Morich and Kalo Marich; *Hindi* : Kali Mirch; *Gujarati* : Kala Mari; *Kashmiri* : Marutis; *Marathi* : Kali Miri and Mire; *Oriya* : Gol Maricha; *Punjabi* : Kali Mirch; *Tamil* : Milagu; *Telugu* : Miriyalu; *Urdu* : Siah Mirch, Kali Mirch; *Sanskrit* : Ushana, Maricha, Hapusha

Black pepper (*Piper nigrum*) is a flowering vine in the family Piperaceae, cultivated for its fruit, which is usually dried and used as a spice and for seasoning. The word 'Pepper' is ultimately derived from the Dravidian word for long pepper, *pippali.* Black pepper is considered the 'King of Spices' as judged from the volume of international trade, being the highest among all the spices known.

2.1. ORIGIN AND DISTRIBUTION

Black pepper is native to Malabar, a region in the western coast of South India; part of the State of Kerala. It is also grown in Malaysia and Indonesia since about that time when it was found in the Malabar Coast. In the last decades of the 20th century, Pepper production increased dramatically as new plantations were founded in Thailand, Vietnam, China and Sri Lanka. The most important producers are India and Indonesia, which together account for about 50% of the whole production volume. Black pepper is mainly cultivated in the southern parts of India comprising of States of Kerala, Karnataka, Tamil Nadu, Andaman and Nicobar Islands, and Pondicherry. India is a major producer, consumer and exporter of Black pepper.

2.2. BOTANICAL DESCRIPTION

- **Plant** - It is an evergreen perennial woody vine growing up to 4 m (13 ft) in height on supporting trees, poles, or trellises. It is a spreading vine, rooting readily where trailing stems touch the ground by means of aerial roots and is not a parasitic.
- **Leaf** -The leaves are oblong, pointed at the tip, alternate, entire, 5–10 cm long and 3–6 cm across.
- **Flower** - The flowers are small, produced on pendulous spikes, 4–8 cm long at the leaf nodes, the spikes lengthening up to 7–15 cm as the fruit matures and mainly hermaphroditic (both sexes in one flower).
- **Fruit** - The fruit of the Black pepper is called a drupe and when dried it is a peppercorn.

The Pepper plant has three types of runners as follows:

1. **Main stem (primary runner)** - The main stems form the permanent stem from which other runners develop.
2. **Secondary runners** - The secondary runners are round, long shoots with lengthy internodes. They climb to a considerable height and later droop downwards.
3. **Tertiary runner** - Tertiary runners are short, sturdier branches that spread horizontally from the axils of primary and secondary runners.

2.3. CLIMATE

Black pepper is a plant of humid region requiring approximately 250 rainy days with a total annual rainfall of 2,000–3,000 mm for its vigorous growth. A dry spell of 30–45 d before flowering is required for successful fruit setting. The hot and humid climate of sub–mountainous tract of Western Ghats is ideal for the cultivation of Black pepper. The crop can tolerate temperatures between10°C and 40°C. Black pepper also requires a high humidity of 75–95 per cent for its healthy growth.

2.4. SOIL

Black pepper can be grown on a wide variety of soil. Virgin soils are ideal for Black pepper cultivation. Soil should be well–drained and rich in organic matter (humus). Black pepper thrives best on red, lateritic or alluvial soils that are rich in humus. A pH of 4.5–6.0 is ideal.

2.5. PROPAGATION

Propagation is usually done by means of cuttings. One or two leaf cuttings are taken only from secondary runners during the month of September. Cuttings are rooted in mist beds and transplanted into the land after 9 months.

Advantages of propagation by cuttings

- Multiplication is rapid.
- The root system is well developed.
- Better field establishment and more vigorous growth as a result of better root system.

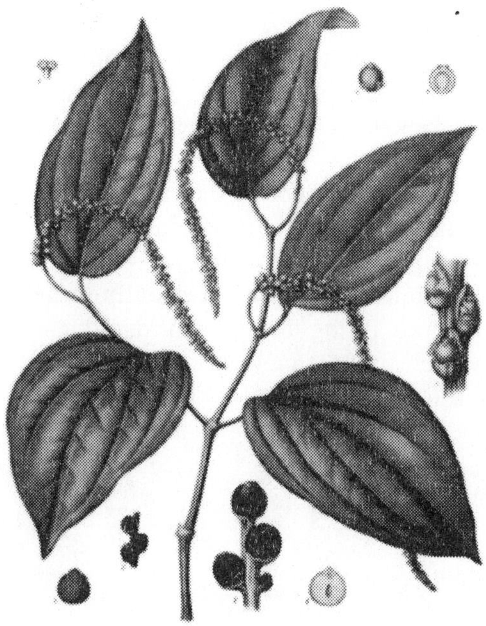

Fig 4. Black pepper

2.6. PLANTING

Ideal time for Black pepper planting is at the onset of southwest monsoons. Rooted cuttings, lifted from the nursery beds or polybags should be transplanted in the main field during morning hours. Two or three rooted cuttings should be planted together on the northern side of the standard or the support tree. Recommended spacing is 2.7 × 2.7 m (row– to–row x plant–to–plant). Recommended plant density in a monocropping system is 1,100 vines/ha, but in a mixed cropping system, it is 540–560 vines/ha. It has a number of medicinal virtues.

2.7. MANURING

Manures

Compost or FYM: 10.0 kg/vine

Fertilizers

The NPK needs to be applied per vine per year as follows:

Nitrogen (N) : 100 g

Phosphorus (P_2O_5) : 40 g

Potassium (K_2O) : 140 g

One-third of the recommended dose should be applied during the first year which should be increased to two-thirds in the second year. Full dose should be given third year onwards. The fertilizers should be applied in two doses, the first half in April with the onset of monsoon and the second half in August–September. The fertilizers should be applied at a distance of about 30 cm all around the vine and at a depth of about 15 cm, and the soil should be forked in after application. Besides, organic manure in the form of cattle manure or compost should be given @ 10 kg/vine in May. Application of lime @ 500 g/vine during April–May in alternate years is also recommended. The supplementation of oil cakes like neem cake (1 kg/vine), composted coir pith (2.5 kg/vine) or composted coffee pulp rich in potassium and suitable microbial cultures of *Azospirillum* and phosphate solubilizing bacteria can improve the fertility.

Fig 5. Panniyur-1, the most popular hybrid Black pepper

2.8. INTERCULTURAL OPERATION

The Pepper vines are tied firmly as and when they grow. The tying should be done around the node, so that the nodal region is firmly attached and pressed against the standards so as to allow the roots to cling to the standards. This is an important operation, which has to be attended carefully otherwise the vine will have no support. The other practice is that when the vine reaches at a height of 75–100 cm without branches, all the leaves should be removed except the terminal 3–4 leaves. The defoliated portions should be covered with friable fine soil. This operations aid in establishing a well–developed root system and lateral branches may arise out of them.

Training and pruning

1. Training and pruning of vines

Training of growing vines is a general practice in Black pepper cultivation. The vines should be trimmed at the top and prevented from growing too tall for convenience of picking. In Tamil Nadu, Pepper vines trained on Silver oak standards are pruned at 6 m height from ground level for easy picking. The vines should be trained up the support to give one main shoot and two lateral orthotropic shoots. These are pruned regularly to encourage the development of lateral fruiting branches but, these lateral branches should not be tied to support, as this would discourage the bushy side growth. The vines are first pruned back to 15–20 cm from the ground level, when they developed 8–9 nodes. Second pruning is done when further 9–10 nodes have been produced, to a height of 3–4 nodes of the previous cut. In this way, vines are pruned 7–9 times until they reach the top of the support about 3 m height. When this is achieved, the terminal growth of vine is arrested by frequent pinching. The lower portion of vine is kept clean and unbranched at least a metre from ground level. Training of vines on the support trees is highly recommended as it facilitates the vigorous growth of the fruit–yielding branches. Removal of unwanted terminal shoots and hanging shoots should be done as and when necessary.

2. Pruning of the standards

Pruning of the standards is also a general practice carried out in a commercial Black pepper garden. This is normally done in March–April every year. Major purpose of this practice is to remove excessive overgrowth of the live standards and to give them a proper shape. The effective height of the standards should be limited to 6 m.

3. Regulation of shade

In Pepper plantation, the Pepper vines should be given shade, especially during the hot weather to keep the soil cool and moist. During cool weather, sunlight should be allowed to encourage production of flowers and fruits. Regulation of shade by lopping the braches of standards is necessary not only for providing optimum light to the vines, but also for enabling the standards to grow straight. Excessive shading during flowering and fruiting encourages pest infestation. Adequate mulch with green leaves, saw dust or coir dust, or organic matter should be given towards the end of the northeast monsoon. To avoid root damage, the base of the vine should not be disturbed. During the second year, practically, the same cultural operations should be repeated. Lopping of standards should be done carefully from fourth year onwards. Usually two diggings are given from the fourth year; one during May–June and the other towards the end of southwest monsoon in October and November. Growing cover crops such as *Calapogonium mucanoides*, *Mimosa invisa* are also recommended under west coast conditions to provide an effective cover to prevent soil erosion during rainy season. Further, these cover crops dry away during summer leaving thick organic mulch.

2.9. HARVESTING

Pepper vines start yielding usually from the third or fourth year. Pepper vines attain full bearing stage in the seventh or eighth year after planting.

The vines flower in May–June. It takes 6–8 months from flowering to ripening stage. Harvesting is done from November–February in the plains and January–March in the hills. When one or two berries on the spike turn red in early, the whole spike is plucked. Harvesting of Pepper is carried out according to the purpose for which it is harvested as follows:

- **White pepper** - To prepare White pepper, the berries are harvested at a slightly advanced stage of ripeness i.e., when the berries turn red (bright orange).
- **Black pepper** - To get Black pepper, the berries are gathered at younger stages.

2.10. PROCESSING

Almost all the produce in India is processed into Black pepper and only a very limited quantity is converted into White pepper.

1. Black pepper

It consists of fully developed, but unripe dried berries of Pepper. The harvested spikes are sun–dried for 7–10 d on cement floor or mats, until the outer skin becomes tough black, shrunk and wrinkled. Drying should be carried till the moisture content gets reduced to 10–15%. Then, the dried berries are separated from the spikes by beating or rubbing between hands or trampling them under the feet. For making good quality of Black pepper of uniform colour, the separated berries are collected in a perforated bamboo basket or vessel and the basket is dipped in boiling water for 1 min. The basket is then taken out and drained. The treated berries are sun–dried on a clean bamboo mat or cement floor. The recovery of black berry is about 33 per cent (26–36 per cent depending upon the variety)

2. White pepper

This consists of dried ripe fruits without pericarp (skin). It is prepared by removing the outer skin along with the pulp before drying. White pepper is prepared by one of the two methods as follows:

(a) Water steeping technique (Traditional method)

(b) Steaming or boiling technique (Improved method)

a. Water steeping technique

It is a traditional and slow method. It involves five steps as follows:

- **Steeping** - Spikes with fully–ripe berries are filled in gunny bags and are steeped in flowing water for about 7–8 d. During this steeping process, the skin gets loosened from the seed.
- **Depulping** - At the end of steeping, the berries are taken out and the skin with the pulp is removed either by rubbing between hands or by trampling under feet.
- **Washing** - These depulped seeds are then washed and cleaned with fresh water repeatedly (3–4 times).
- **Drying** - The cleaned seeds are sun–dried for 3–5 d on cement floor or mats till they become white and the moisture gets reduced to 10–15 per cent.
- **Polishing** - The dried seeds are now dull white in colour. They are further cleaned by winnowing or by rubbing with a cloth. The percentage of recovery of White pepper is about 25 per cent of ripe berries.

b. Steaming or boiling technique

This is an improved and quick method developed at CFTRI, Mysore. It involves four steps as follows:

- **Boiling** - Freshly harvested spikes or berries are boiled for about 15 min.
- **Depulping** - The boiled berries are then pulped mechanically. Boiled berries pass through motorized fruit pulping machine.
- **Bleaching** - The deplumed berries are washed thoroughly by using bleaching powder or any bleaching agent.
- **Drying** - The cleaned berries are sun–dried for 3–5 d on cement floor or mats till they become white and the moisture gets reduced to 10–15 per cent.

2.11. YIELD

A full–bearing vine yields 1 kg of dry Pepper. Seven or eight year old Pepper vine gives 800–1,000 kg/ha of Black pepper.

2.12. USES

Black pepper is produced from the still–green unripe berries of the Pepper plant. Different parts of Black pepper are used for different purposes. Black pepper contains several medicinal properties. The fruits are analgesic, antiseptic, antispasmodic, diaphoretic, febrifuge, rubefacient, stimulant, stomachic, acrid, bitter, carminative, aphrodisiac, diuretic and digestive. Black pepper is aromatic and has slightly musty odour. The essential oil of Black pepper is found largely in its flesh and skin, and its pungent taste comes from the alkaloids. The resins are found mostly in seeds of Black pepper.

Black pepper is most commonly used for treating illnesses including constipation, diarrhoea, earache, gangrene, heart disease, hernia, hoarseness, indigestion, insect bites, insomnia, joint pain, liver problems, lung diseases, oral abscesses, sunburn, tooth decay, toothaches, anemia, poor blood circulation, jaundice etc.

Table 17. Major Black pepper cultivars available in India

Name of cultivars	Important characters
Aimpiryan	High yielder, high fruit set, excellent in higher elevations, good in quality.
Arakkulammunda	Moderate and regular but early bearer, medium in quality.
Balankotta	Cultivar with large drooping leaves, moderate but early irregular bearing, medium in quality, bold berries.
Bilimallegesara	Moderate yielder, good for State of Karnataka.
Chengannurkodi	Moderate yielder from South Kerala, medium in quality.
Cheppukulamundi	Moderate yielder from Central Kerala, medium in quality.
Cheriyakaniakadan	Popular in North Kerala, moderate and early bearing variety.
Jeerakamundi	Cultivar with small leaves, short spikes with high spiking intensity, and small berries; alternate bearing variety.
Kalluvally	A promising cultivar from North Kerala, good yielder, medium in quality, high dry recovery, drought tolerant.
Karimunda	Most popular cultivar, suitable for most of the Black pepper growing areas, high yielder, regular bearer, medium in quality.
Kottan	A cultivar found in North Kerala, moderate in yield, medium in quality
Kottanadan	A high yielding, high quality cultivar from South Kerala, drought tolerant and regular bearer.

Kurimalai	A cultivar from Karnataka, moderate yielder, medium in quality
Kuthiravally	A cultivar with long spikes, high yield and good quality.
Kuttianikodi	A moderate yielder from Central Kerala with relatively long spikes and good spiking intensity.
Malamundi	A moderate yielder, medium in quality
Malligesara	A common cultivar from Karnataka, relatively good in yield.
Manjamundi	A moderate yielder from North Kerala, medium in quality.
Narayakodi	Popular in South Kerala, moderate yielder, medium in quality.
Neelamundi	A good yielder from Central Kerala, medium in quality, tolerant to *Phytophthora.*
Nedumchola	A cultivar with small leaves and short spikes, poor yielder.
Neyyattinkaramundi	A cultivar from Central Kerala, medium in quality and yield.
Perambramunda	A cultivar from North Kerala, moderate yielder, medium quality.
Perumkodi	A cultivar from Central Kerala, moderate in yield and quality.
Poonjaranmunda	A cultivar originally from Central Kerala, sporadically found in gardens of North Kerala, moderately good in yield with long spikes and quality.
Thommankodi	A cultivar from Central Kerala, good yielder and quality.
Thulamundi	A Central Kerala cultivar, medium in yield and quality.
Uddagara	A popular cultivar of Karnataka, good in yield, medium in quality.
Vadakkan	A triploid cultivar from North Kerala, medium in quality and yield, very large berries.
Valiyakaniyakadan	A cultivar with larger leaves, medium in yield and quality.
Vattamundi	A moderate yielder from Central Kerala.
Vellanamban	Relatively moderate yielder, medium in quality, characterized by the white colour of the young shoot tip.

Source: Botany and Crop Improvement of Black Pepper by Ravindran (2000)

Fig 6. Subhakara, a high–yielding variety of Black pepper

Table 18. Released Black pepper varieties and their characteristics

Variety	Average yield (kg/ ha)	Drainage (%)	Quality attribute			Remarks
			Piperine	Oleoresin	Essential oil	
Panniyur 1 (Krishna) (Hybrid between Uthirankotta x Cheriyakaniyakadan)	1,242	35.3	3.5	11.8	2.4	Long spikes with large berries, high in oleoresin, early bearing, performs well under open situations, suitable to all Pepper–growing regions, not suited to heavily–shaded areas.
Panniyur 2 (Open pollinated progeny of Balankotta)	2,570	35.7	10.9	3.4	5.2	Shade tolerant, rich in oleoresin and piperine, suited to all Pepper–growing tracts of Kerala.
Panniyur 3 (Shima) (Hybrid between Uthirankotta x Cheriyakaniyakadan)	1,953	27.8	12.7	3.1	3.2	Vigorous and late maturing, suitable for all Pepper–growing region, performs well under open situation, long spikes and bold berries.
Panniyur 4 (Clonal selection from Kuth Panniyur 2, Open pollinated progeny of Balankottairvally)	1,277	34.7	4.4	9.2	2.1	Stable yielder, performs well under adverse condition including partial shade.
Panniyur 5 (Open pollinated progeny of Perumkodi)	1,098	35.7	5.3	12.3	3.8	Long spikes, shade tolerant variety, better suited for mixed cropping in coconut/ arecanut gardens, tolerant to nursery diseases.
PLD-2 (Selection from Kottanadan)	2475	-	3.3	15.5	3.5	High oleoresin recommended for Trivandrum and Quilon districts of Kerala.

Subhakara (Selection from Karimunda)	2352	35.5	3.4	12.4	6.0	Adaptable to all Pepper–growing tracts, high quality line with high oleo-resin and essential oil.
Sreekara (Selection from Karimunda)	2677	35.0	5.1	13.0	7.0	Adaptable to all Pepper–growing tracts, high quality line with high oleo-resin and essential oil.
Panchami (Selection from Aimpiriyan)	2828	34.0	4.7	12.5	3.4	Late–maturing variety with excellent fruit set
Pournami (Selection from Germplasm)	2333	31.0	4.1	13.8	3.4	High–yielding variety, tolerant to root knot nematode, suited to all Pepper–growing regions of Kerala.
Panniyur 6 (Clonal selection from Karimunda)	2127	33.0	4.9	8.3	1.3	Steady and stable yielder, tolerant to drought and adverse climatic conditions, suitable for open condition as well as partial shade.
Panniyur 7 (Open pollinated progeny of Kalluvally)	1410	33.6	5.6	10.6	1.5	Vigorous, hardy and a regular bearer, long spike, high piperine, tolerates adverse climatic conditions, suitable for open and shaded conditions.

Source: Vegetable Crop: Breeding of Spice Crop by K.V. Peter, P. N. Rabindran and K. Nirmal Babu (2007)

Table 19. Improved varieties of Black pepper and their parentage and yield

Name of hybrid / variety	Parentage	Yield / vine (kg)
Panniyur-1	F1 hybrid Uthirankotah x Cheriyakaniakadan	2.5
Panniyur-2 (Krishna)	Open pollinated seedlings of Balankottah	4.5
Panniyur-3 (Shima)	F1 hybrid of Uthirankotah x Cheriyakaniakadan	4.4
Panniyur-4	Selection from Kuth Panniyur-2	2.3
Panniyur-5	Open pollinated progeny of Perum kodi	2.75

Sreekara	Selection from Kanmundu (KS 14)	4.8
Subhakara	Selection from Kariamunda (KS 27)	4.2
Panchami	Selecton from Aimpiriyan Coll. 856	5.2
Pournami	Selection from Ottaplackal Type coll. No. 812	4.7

2.13. PLANT PROTECTION

Insect Pests

1. Pollu beetle

Scientific name - *Longitarsus nigipennis*

The pollu beetle (*Longitarsus nigripennis*) is the most destructive pest of Black pepper and is more serious in plains and at altitudes below 300 m.

Nature of damage

The adult beetles feed on tender leaves and spikes and damage them. The females lay eggs on tender spikes and berries. The grubs bore into and feed on the internal tissues, and the infested spikes turn black and decay. The infested berries also turn black and crumble when pressed.

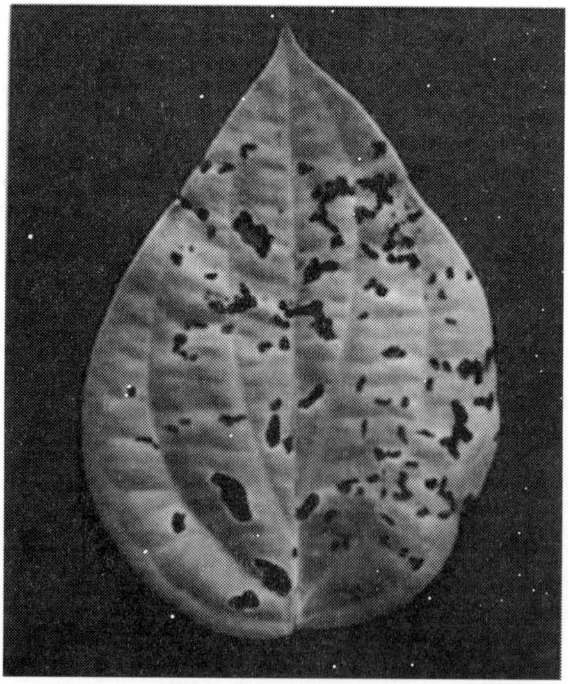

Fig 7. Pollu–affected leaf

Control measures

The insect can be controlled by spraying any one of the insecticides in the Table 20 as soon as the incidence of the insect is noticed.

Table 20. Insecticides used for controlling Pollu beetle of Black pepper

Sr. No.	Name of insecticide	Amount (per litre of water)
1.	Carbosulfan 25 EC	2.0 ml
2.	Acephate 75 WP	0.75 g
3.	Thiomethoxam 25 WG	0.2 g
4.	Lamdacyhalothin 25 EC	0.75 ml

2. Leaf gall thrips

Scientific name - *Liothrips karnyi*

The leaf gall infestation is generally serious at higher altitude and also in nurseries in the plains. Many of the vines infested by the bugs are often attacked by a fungal pathogen called *Phytophthora capsici* and plant parasitic nematodes such as *Meloidogyne incognita* and *Radopholus similes,* which cause the death of the vines.

Nature of damage

The leaf gall thrips infest the leaves of Black pepper and induce the formation of marginal, tubular galls on the leaves.

Control measures

The insect can be controlled by spraying any one of the insecticides given in Table 21 as soon as the incidence of the insect is noticed.

Table 21. Insecticides used for controlling the Leaf gall thrips of Black pepper

Sr. No.	Name of insecticide	Amount (per litre of water)
1.	Carbosulfan 25 EC	2.0 ml
2.	Acephate 75 WP	0.75 g
3.	Thiomethoxam 25 WG	0.2 g
4.	Clothianidin 50 WDG	0.1 g
5.	Lamdacyhalothin 25 EC	0.75 ml

3. Scale insect

Scientific name- Mussel scale - *Lepidosaphes piperis*

Coconut scale - *Aspidiotus destructo*

The scale insect sometimes causes serious damage to Black pepper especially in higher altitude and also to older cuttings in nurseries in the plains.

Nature of damage

The scale insect encrust stems, leaves and berries and suck the sap from the affected parts of the plant and cause yellowing and wilting of infested portions; in severe cases of infestation, the affected portions of vines dry up especially in summer.

The insect pest infestation is more severe during the post–monsoon and summer periods.

Control measures

- The severely infested branches should be clipped off and destroyed.
- The insect pest can be controlled by spraying any one of the insecticides in the Table 22.

Table 22. Insecticides used for controlling the Scale insect of Black pepper

Sr. No.	Name of insecticide	Amount (per litre of water)
1.	Chloropyriphos 25 EC	2.5 ml
2.	Carbosulfan 25 EC	2.0 ml
3.	Acephate 75 WP	0.75 g
4.	Imidachloroprid b17.8 SL	0.2 ml

4. Top shoot borer

Scientific name - *Cydia hemidoxa*

The top shoot borer (*Cydia hemidoxa*) is a serious pest in younger plantations in all Black pepper crops. The pest infestation is higher during July–October when numerous succulent shoots are available in the vines.

Nature of damage

The larvae bore into tender terminal shoots and feed on internal tissues resulting in blackening and decaying of affected shoots and ultimately affecting the growth of vine.

Control measures

The insect pest can be controlled by spraying any one of the insecticides given in Table 23.

Table 23. Insecticides used for controlling the Top shoot borer of Black pepper

Sr. No.	Name of insecticide	Amount (per litre of water)
1.	Chloropyriphos 25 EC	2.5 ml
2.	Carbosulfan 25 EC	2.0 ml
3.	Acephate 75 WP	0.75 g
4.	Imidachloroprid 17.8 SL	0.2 ml

5. Mealy bug

Scientific name - *Planococcus minor* (Maskell)

The infestation of mealy bug is becoming increasingly severe in recent years, especially at higher altitudes of Wayanad district in Kerala and Kodagu district in Karnataka. The infestation is generally more during the post–monsoon season and lesser during summer months. The infestation is often noticed on nursery plants and also on plants of all ages in the field.

Nature of damage

Mealy bug affects the aerial parts of the Black pepper vines such as the tender shoots, leaves and berries, and sucks the sap from the affected parts. Infestation of mealy bugs leads to yellowing, wilting of leaves and defoliation of younger vines, which succumb to the pest attack during the course of 2–3 yr of planting.

Control measures

- Handpicking and killing the insect, if possible, is found effective in controlling the mealy bug. This method is applicable for controlling the insect in small area.
- Lady bird beetle such as *Cryptolisus montochari* can be used for controlling the Mealy bug.
- The insect can be controlled by spraying simple water on the insect–affected plants.
- The insect pest can be controlled by spraying any one of the insecticides in the Table 24.

Table 24. Insecticides used for controlling the Mealy bug of Black pepper

Sr. No.	Name of insecticide	Amount (Per litre of water)
1.	Chloropyriphos 25 EC	2.5 ml
2.	Carbosulfan 25 EC	2.0 ml
3.	Acephate 75 WP	0.75 g
4.	Imidachloroprid 17.8 SL	0.2 ml

Diseases

1. Foot rot

C.O.: *Phytophthera capsici*

Phytophthora infections are noticed on leaves, stems and roots of cuttings in the nursery.

Symptoms

Dark spots with fimbriate margins appear on the leaves, which spread rapidly resulting in defoliation. The infections on the stem are seen as black lesions, which result in blight. The symptoms on the roots appear as rotting of the entire root system.

Control measures

The disease can be controlled by spraying any one of the fungicides given in Table 25.

Table 25. Fungicides used for controlling the Foot rot of Black pepper

Sr. No.	Name of fungicide	Amount (per litre of water)
1.	Coper oxychloride 50 WP	5.0 g
2.	Carbendazim 50 WP	1.0 g
3.	Mancozeb 75 WP	2.5 g
4.	Chlorothalonil 75 WP	2.0 g
5	Trichoderma viridi	4.0 g

2. Anthracnose (Pollu disease)

C.O.: *Colletotrichum gloeosporioides*

The fungus infects the leaves causing yellowish brown to dark brown irregular leaf spots with a chlorotic halo. The affected berries show brown sunken patches during early stages, affecting their further development. In later stages, the discolouration gradually increases and the berries show the characteristic cross–splitting. Finally, the berries turn black and dry.

Control measures

The disease can be controlled by spraying any one of the fungicides given in Table 26.

Table 26. Fungicides used for controlling the Anthracnose of Black pepper

Sr. No.	Name of fungicide	Amount (per litre of water)
1.	Dithiolone 75 WP	2.5 g
2.	Thiophanate Methyl 70 WP	1.0 g
3.	Dodaine 65 WP	0.5 g

3. Leaf rot and blight

C.O.: *Rhizoctonia solani*

Leaf rot and blight is often serious disease in nurseries during April–May when warm humid conditions prevail.

Symptoms

The fungus infects both leaves and stems.

- Symptoms on leaf - Grey sunken spots and mycelia threads appear on the leaves and the infected leaves are attached to one another with the mycelia threads.
- Symptoms on stems - The infection occurs on stem as dark brown lesions, which spreads both upwards and downwards. The new flushes subtending the points of infection gradually droop and dry up.
- Leaf spots causedby *Colletotrichum* species are characterized by yellow halo surrounding.

Control measures

The disease can be controlled by spraying any one of the fungicides given in Table 27.

Table 27. Fungicides used for controlling the Leaf rot and blight of Black pepper

Sr. No.	Name of fungicide	Amount (per litre of water)
1.	Coper oxychloride 50 WP	5.0 g
2.	Carbendazim 50WP	1.0 g
3.	Mancozeb75 WP	2.5 g
4.	Chlorothalonil 75 WP	2.0 g

4. Basal wilt

C.O.: *Sclerotium rolfsii*

The disease is mainly noticed in nurseries during June–September.

Symptoms

Grey lesions appear on stems and leaves. White mycelium are seen on the leaves at the advancing edges of the lesions. The mycelia threads later girdle the stem resulting in drooping of leaves beyond the point of infection and in advanced stages, drying of the rooted cuttings. Small, whitish to cream coloured, grain–like sclerotic bodies appear on the mature lesions.

Control measures

The disease can be controlled by spraying any one of the fungicides given in Table 28.

Table 28. Fungicides used for controlling the Basal wilt of Black pepper

Sr. No.	Name of fungicide	Amount (per litre of water)
1.	Mancozeb 75 WP	2.5 g
2.	Coper oxychloride 50 WP	5.0 g
3.	Zineb 75 WP	2.5 g
4.	Copper hydroxide 77 WP	2.0 g
5.	Dipheniconazol 25 EC	1.0 ml
6	*Trichoderma viridi*	5.0 g.

3 GINGER

Botanical name - *Zingiber officinale*

Family - Zingiberaceae

Commercial part - Rhizome

Indian names - *Bengali* : Ada; *Hindi* : Adrak; *Gujarati* : Adu; *Kannada* : Shunti, Ardraka; *Malayalam* : Inchi; *Marathi* : Ale; *Oriya* : Ada; *Punjabi* : Adrak; *Sanskrit* : Ardraka; *Tamil* : Inji; *Telugu* : Allamu, Sonthi; *Urdu* : Adrak, Adhrak.

Ginger *(Zingiber officinale* Roscoe) is a fibrous–rooted perennial plant with branched underground stems called rhizome. The rhizomes are used as spice and medicine.

Ginger or Ginger root is the rhizome of the plant *Zingiber officinale*, consumed as a delicacy, medicine, or spice. Ginger is the third most–important spice originated in South Asia. Ginger (*Zingiber officinale*) is an ancient and major spice of the world. Ginger (*Zingiber officinale*) (Family: *Zingiberaceae*) is an herbaceous perennial, the rhizomes of which are used as a spice. India is a leading producer of Ginger in the world.

The English name 'Ginger' comes from French word *gingembre*, Old English: *gingifere*, Medieval Latin: *ginginer*, Greek: *zingíberis*. Ultimately, the origin is from Tamil word 'inji ver' or Malayalam word 'inji veru'. The botanical term for root in Tamil is *ver* and Malayalam is *veru*, hence inji root or injiver in Orissa.

3.1. ORIGIN AND DISTRIBUTION

Ginger is a native of tropical south or Southeast Asia (Baily 1961 and Purseglove 1972) and now cultivated in India. It was brought to Mediterranean region from India by traders during first century A.D. During the 13th century A.D., the Arabs took Ginger to East Africa from India. Later, it was spread to West Africa by Portuguese for commercial cultivation. It is mainly grown in tropical and subtropical countries such as India, China, Taiwan, Philippines,

Sierra–Leone, Jamaica, Fiji, Mexico, Brazil and Nigeria on commercial scale. Ginger is cultivated in Indian States such as Andhra Pradesh, Arunachal Pradesh, Bihar, Chhattisgarh, Gujarat, Haryana, Himachal Pradesh, Jammu and Kashmir, Karnataka, Kerala, Meghalaya, Orissa, Rajasthan, Tamil Nadu and West Bengal. India is the largest producer of dry Ginger in the world contributing to the tune of about 30 per cent of the world's production.

3.2. BOTANICAL DESCRIPTION

- **Plant** - It is an herbaceous perennial plant (Fig. 8) having underground branched rhizome with small scales. The inner core of the rhizome is pale yellow to bluish tinge while the outer is light yellow in colour. Adventitious roots and storage roots arise from the nodes of these scales. The ancillary buds, which shoot up as leafy stem (known as pseudo stem), die out annually but the plant continues to live through its rhizome.
- **Leaf** - The leaves are sheathing, arranged alternatively, linear lanceolate, gradually acuminate and glabrous.
- **Flower** - The flowers are borne on a spike produced in a peduncle different from the aerial leafy stem, arising directly from the rhizome. The flowers are many, trimerous bisexual, irregular, epigenous, yellow in colour with dark purplish spots.
- **Fruit** -The fruit is very rarely produced, which is an oblong capsule. (Ravindran and Babu 2005, Purse glove 1981), but *Zingiber officinale* is not known to set seeds.

3.3. CLIMATE

Ginger grows best in warm and humid climate. It is grown as rain fed and irrigated crop. The prerequisite of climate for Ginger is as follows:

Rainfall (Annual)	: 125–250 cm
Temperature	: 28°C–35°C
Altitude	: 1,500 m

3.4. SOIL

Ginger grows in different kinds of soil. But deep, well–drained friable loam, rich in humus is ideal for the cultivation of Ginger. Ginger thrives best in light soil such as sandy, sandy loam, loam etc.

3.5. PROPAGATION

Clonal multiplication of Ginger from vegetative buds has been reported by many workers (Hosoki and Sagawa 1977, Nadgauda et al., 1980, Nirmal Babu et al., 1997, Sharma and Singh 1997, Rout et al., 2001). Diseases of Ginger are often spread through infected seed rhizomes. Tissue culture technique would help in the production of pathogen–free planting material of elite varieties. Field evaluation of tissue–cultured plants indicated that it required at least two crop seasons to develop rhizomes of normal size that can be used as seed rhizomes for commercial cultivation. Molecular characterization of micropropagated plants indicated genetic uniformity (Rout et al., 1998) as well as certain polymorphism (Nirmal Babu et al., 2003).

Fig. 8. Ginger

3.6. PLANTING

1. Time of planting

Rainfed condition - The best time for planting Ginger in the west coast of India is during the first fortnight of May with the receipt of pre–monsoon showers.

Irrigated condition - The best suited time for planting is the middle of February or early March (for vegetable Ginger).

2. Method of planting

The rhizome bit of 20–25 g weight is planted in small pits at a spacing of 20 x 20 cm to 25 x 25 cm and at a depth of 4–5 cm with at least one viable healthy bud facing upwards. The first mulching should be done at the time of planting with quick rotting green leaves at the rate of 10–12 t/ha or with dry leaves at the rate 5–6 t/ha. Mulching is essential as it enhances sprouting, and increases infiltration and organic matter. Green manuring crops such as dhaincha and sannhemp may be grown in the interspaces of beds along with Ginger and green manure crop may be harvested during the second mulching of Ginger beds.

3. Seed treatment

The seed materials should be treated with any one of the fungicides/organic fungicide in the Table 29, before planting:

Table 29. Fungicides/organic fungicides used for seed treatment of Ginger

Sr. No.	Name of fungicide/bio-fungicide	Amount (per litre of water)
1.	Carbendazim 50 WP	2.0–3.0 g
2.	M.E.M.C 6 WP	1.0 g
3.	Captan 50 WP	2.0–5.0 g
4.	Mancozeb 75 WP	2.5 g
5.	Trichoderma viridi	4.0–5.0 g

The treated seeds should be kept under shade for at least 24 h and then planted.

4. Seed rate

1,500.0 kg/ha

3.7. MANURING

Manures

Compost or FYM: 25.0–30.0 ton/ha

Fertilizers

Nitrogen (N)	: 100.0–120.0 kg/ha
Phosphorus (P_2O_5)	: 75.0–80.0 kg/ha
Potassium (K_2O)	: 100.0–120.0 kg/ha
Sulphur (S)	: 20.0–250 kg/ha

Sulphur should be applied at the time of land preparation to correct the deficiency of Sulphur. The deficiency of Sulphur is increasing in the Indian soils.

Method of application

Manures - Manures should be applied as basal dressing at the time of land preparation and well mixed by ploughing followed by planking.

Basal dressing

Fertilizers - Half quantity of nitrogen and potassium and entire quantity of phosphorus should be applied as basal dressing.

Top dressing

The remaining quantity of nitrogen and potassium should be split in two doses as top dressing at the 45 and 90 DAP (Days after planting).

3.8. IRRIGATION

The critical stages of irrigation are sprouting, rhizome initiation and rhizome development stage. The crop rises in the month of April–May, needs 2–4 initial watering at an interval of seven days depending upon the soil types. After this, the crop should receive monsoon rain and it will then come up well till the end of September. Subsequently, the crop has to be given watering commencing from middle of October and the end of December at 15 d intervals.

3.9. INTERCULTURAL OPERATION

1. Weeding

The field should be kept free from weeds. For this, two weedings should generally be given to the Ginger crop. First weeding should be done just before the second mulching. It should be repeated depending on the intensity of the weed growth or at an interval of 45–60 d. During hoeing, every care should be taken so that the rhizomes should not be disturbed, injured or exposed.

2. Earthing up

Earthing up is essential to prevent exposure of rhizomes and provide sufficient soil volume for free development of rhizomes. It should be done at 45 and 90 DAP, immediately after weeding and application of fertilizers.

3. Mulching

Mulching with green leaves or organic wastes is essential to prevent soil splashing and erosion of soil due to heavy rain. It also adds organic matter to the soil, checks weed emergence and conserves moisture during the latter part of the cropping season. The first mulching should be done at the time of planting with green leaves @ 10–12 ton/ha. Mulching should be repeated @ 7.5 ton/ha at 45 and 90 DAP, immediately after weeding, application of fertilizers and earthing up.

1. Harvesting

- **Fresh Ginger** - The fresh Ginger should be harvested before attaining the full maturity i.e., when rhizomes are still tender, low in pungency and fibre content, usually from fifth month onwards after planting.
- **Mature Ginger** - The mature Ginger should be harvested after 5–7 months of planting while harvest for dried spices and oil is best at full maturity i.e., between 8–9 months after planting when leaves start yellowing. Rhizomes to be used for planting material should be harvested until the leaves become completely dry.

2. Cleaning of rhizome

After harvesting, the fibrous roots attached to the rhizome should be trimmed off and soil should be removed by washing. Rhizomes should be soaked in water overnight and then cleaned. The skin can be removed by scrapping with the correct instrument. Peeling or scraping reduces drying time, thus minimizing mold growth and fermentation. The cured rhizomes should be treated with fungicides such as Mancozeb 50 WP @ 3–4 gm per litre of water, dried in shade, and stored in pits covered with 20 cm layer of sand alternating every 30 cm layer of rhizomes. These pits should be dug under a thatched roof to protect the rhizomes from rain, water and direct sunlight.

3.10. YIELD

Average yield varies from 12–15 ton/ha. However, recovery of dry Ginger varies from 20–22 per cent.

Table 30. Cultivar diversity of Ginger in India

Cultivar	Special features of rhizome
Assam	Bold, highly flavoured, pungent and highly fibrous
Burdwan-1	Bold, pungent, flavoured and fibrous
Ernad Chernad	Bold, highly pungent, flavoured and fibrous
Gurubathan	Bold, buffed skin, flavoured, pungent and fibrous
Himachal Pradesh	Bold, lemon–flavoured and fibrous
Jorhat	Bold, pungent, highly flavoured and fibrous
Karakkal	Bold, buffed skin, pungent, flavoured and less fibrous
Kuruppampadi	Bold, pungent, flavoured and fibrous
Mananthodi	Bold, pungent, flavoured and fibrous

Maran	Bold, pungent, flavoured and fibrous
Nadia	Slender, pungent, lemon flavoured and less fibrous
Narasapattam	Bold, buffed skin, pungent, flavoured and less fibrous
Poona	Bold, buff skin, flavoured and less fibrous
Taffingiva	Extra bold, yellowish white skin, pungent, flavoured and less fibrous
Thinladium	Bold, buff skin, pungent, flavoured and fibrous
Thingpuri	Slender, skin whitish gray, slightly pungent, flavoured and less fibrous
Tura	Slender, pungent, flavoured and fibrous
Uttar Pradesh	Bold, buff skin, pungent, flavoured and fibrous
Valluvanad	Slender, pungent, flavoured and fibrous
Wynad local	Bold, buff skin, pungent, flavoured and less fibrous
Wynad Kunnamangalam	Bold, pungent and flavoured

Source: Advances in Horticulture, 1994

Table 31. Particulars about the variety of Ginger

Character/ trait	Variety /Cultivar/ Accession	References
High yield	UP, Rio–de–Janeiro, Thingpuri, Karakkal, Suprabha, Anamika, Jugijan SG–646, SG–666	Mohanty and Panda (1994) Rattan (1989)
	Rio-de-Janeiro, Suprabha, Suruchi, Suravi, Jugijan, Thingpuri, Wynad Local, Himachal, Karakkal, Varada, Maran, Acc. Nos. 64, 117 and 35 Rio–de–Janeiro, Maran, Nadia, Narasapattam Rio–de–Janeiro, China, Ernad Chernad	Sasikumar et al (1994) Sarma et al (2001) Paulose (1973) Thomas (1966)
	Wynad, SG-700, SG-705 and BDJR-1226, V_2E_4-5, PGS-43, SG-876, SG-882	AICRPS (1999, 2000, 2001)
Bold rhizome	China, Taffingiva, SG-35	Sasikumar et al (1994)
Slender rhizome	Varada, Gurubathan, Bhaise, China,	Sasikumar et al (1999)
Short duration	Acc. Nos. 117, 35, 15, 27 and 142	Sarma et al (2001)
	Suruchi, Kunduli local	Mohanty and Panda (1994)
	Sierra-Leone	Mohanty and Panda (1994)

Source: Advances in Spice Research, 2006

Table 32. Promising selections of Ginger for quality attributes

Character/ trait	Variety /Cultivar/ Accession	References
High dry recovery (%)	Tura local, Thodupuzha, Nadia, Kuruppampadi, Maran, Thinladium, Jorhat, Vengara, Ernad Chernad, Himachal Pradesh, Sierra-Leone, Suruchi, Assam, China, Mowshom, Thingpui, Varada, Acc. Nos. 27, 117, 204 and 294, SG–685, Narasapattam, Rio–de–Janeiro	Mohanty (1984); Sreekumar et al (1980); Nybe et al (1980); Nair et al (1969); Muralidharan (1972); Thomas (1966); Sasikumar et al (1994 and 1999) Paulose (1973)
High Oleoresin (%)	Assam, Mananthody, Wynad Local Kuruppampadi, Rio–de–Janeiro, Maran, Wynad Kunnamangalam, Ambalavayalan, Santhing Pui, Himachal, Varada, China, Santhing Pin (Manipur-I), Ernad Chernad, Erattupetta, Tamarassery local, PGS-33 and PGS-1, Nadan Pulpally, Nadan Acc. Nos. 3, 6, 14, 57, 110, 118, 582, 236, 388, 414	Krishnamoorthy et al (1970) Natarajan et al (1972) Muralidharan (1972) Nybe et al (1980); Sreekumar et al (1980); Sasikumar et al (1994, 1999); Sarma and Sasikumar et al (2002) Zachariah et al (1993, 1999)
High Essential oil (%)	V_1S_1-8, BDJR–1226, Chanog–II, Mananthody, Karakkal, Rio–de–Janeiro, Vengara, Valluvanad, Elakallan, Sabarimala, Acc. Nos. 118, 14,64,418, 399, 389, 205, 110, 236, 104, 296, Nadan Pulpally, Thodupuzha, BLP-6, SG-723, BDJR-1054, SG-55, Maran, Shilli, Bangi, Himgiri, Acc. No. 389, V_1E_4-4, PGS-23 and SG-706	Krishnamurthy et al (1970) Lewis et al (1972) Nybe et al (1980) Sarma et al (2001) Sarma and Sasikumar (2002) Zachariah et al (1999)
Low crude fibre (%)	China, Uttar Pradesh, Himachal Pradesh, Nadia, Tura, Ernad–Chernad, Zahirabad, Kuruppampadi, Mizo, PGS-16, Jamaica, Acc. Nos.15, 27, 287, 288, 22, 18, 419, 386, 415, 200, 110, 336, Varada	Thomas (1966); Sreekumar et al (1980); Nybe et al (1980); Sasikumar et al (1994, 1999); Sarma and Sasikumar et al (2002) Zachariah et al (1999)
High Gingerol and Shogaol	Wynad Kunnamangalam, Ambalavayalan, Ernad Chernad, Sawthing Pui, Rio–de–Janeiro, Mizo, Nadia, Maran, Ernad Chernad, Kada, Narianpara, Santhing Pin (Manipur–I), PGS–37, S–641, Maran, Erattupetta, Nadan Pulpally, Jorhat Local, PGS–16, Baharica and Kunduli	Sasikumar et al (1994,1999) Sarma et al (2001) Zachariah et al (1993)
High zingiberene and (6) Gingerol	Baharica and Amaravathy	Sasikumar et al (1999)
Salted Ginger	Acc. Nos. 35, 27	Sasikumar et al (1999)

Source: Advances in Spice Research, 2006

Table 33. List of Ginger germplasm tolerant to insect pest and diseases

Character	Cultivar/ Accession	Reaction References
Insect - Pests	Rio-de-Janeiro	Tolerant - Nybe et al (1980)
Shoot borer	Wild-2	Least infestation - Mohanty (1984)
Rhizome scale	Anamika	Sasikumar et al (1994)
Storage pest	Varada, Acc. Nos. 215 and 212, Valluvanad, Tura, Himachal Pradesh, Acc. Nos. 36, 59 and 221	Resistant - Sarma et al (2001)
Root–knot nematode		Least infestation - Charles and Kuriyan (1980)
Diseases		Resistant - Sarma et al (2001)
Rhizome rot (*Pythium* species)	Jorhat and Sierra-Leone	Least incidence - (11.25%) AICSCIP (1975a)
Bacterial wilt (*Ralstonia solanacearum*)	Maran	Least infection - Nybe and Nair (1979)
	Narasapattam	Least susceptible (1–20%) - Mohanty (1984)
Leaf spot (*Phyllosticta zingiberi*)	Burdwan-1, Anamika, Poona and Himachal	Less susceptible - Sasikumar et al (1994)
	BDJR-1226, Jamaica, BLP-6	Less susceptible - AICRPS (1999)
	V_2E_5-2, Rio–de–Janeiro	Least incidence - Pradeep Kumar et al (2000)
	Taffingiva, Bajpai, Nadia, Maran and Kunduli local	Most tolerant - Nybe and Nair (1979)
		Less susceptible - Sasikumar et al (1994)

Source: Advances in Spices Research, 2006

Fig 9. Ginger

Varieties of Ginger according to some quality

High yielding types	: Maran, Karakkal, Rio de Janeiro, Mahim
Less fibre content	: Jamaica, Bangkok, China Thingpuri (Exotic good for Green Ginger)
High oleoresin	: Emad Chemed, China, Karuppamadi, Rio–de–Jeneiro
High dry Ginger recovery	: Karakkal, Nadia, Maran, Tura
High volatile oil	: Sleeva local, Narasapattam, Emad, Chemad

Table 34. Released varieties of Ginger and their salient features

Cultivar	Institution/ Released from	Average Yield (fresh t/ha)	Maturity (days)	Dry recovery (%)	Crude recovery (%)	Oleoresin (%)	Remarks
Suprabha	Orissa University of Agriculture & Technology, Pottangi	16.6	329	20.5	4.4	8.9	Plumpy rhizome, wider adaptability, suitable for both early & late sowing
Suruchi	Orissa University of Agriculture and Tech. Pottangi	11.6	218	23.5	3.8	10.0	Bold rhizome, suitable for rainfed and irrigated conditions
Suravi	Orissa University of Agriculture and Tech. Pottangi	17.5	225	23.0	4.0	10.2	Plumpy rhizome, suitable for rainfed and irrigated conditions
Himgiri	Y.S. Parmar University of Horticulture and Forestry, Solan, H.P.	13.14	230	20.2	6.05	4.29	Best for Green Ginger, less susceptible to rhizome rot, suitable for rainfed condition
IISR-Varada	Indian Institute of Spices Research, Calicut, (Ker.)	22.6	200	20.7	3.29–4.5	6.73	Bold rhizome, tolerant to rhizome rot, wide adaptability
IISR-Mahima	Indian Institute of Spices Research, Calicut, (Ker.)	23.2	300	23.0	3.26	4.48	Plumpy, bold rhizome with low fibre content
IISR-Rejatha	Indian Institute of Spices Research, Calicut, (Ker.)	22.4	200	20.8	4.0	6.34	Plumpy, round and bold rhizome with low fibre and high oil content.

Source: Vegetable Crop: Breeding of Spice Crop by K.V. Peter, P. N. Rabindran and K. Nirmal Babu (2007)

3.11. HARVESTING

The crop is ready for harvest in about eight months after planting when the leaves turn yellow, and start drying up gradually. The clumps should be lifted carefully with a spade or digging fork, and the rhizomes should be separated from the dried–up leaves, roots and adhering soil. The Ginger should be harvested according to its uses as follows:

Vegetable Ginger

The harvesting of vegetable Ginger is done from sixth months onwards. The rhizomes are thoroughly washed in water and sun–dried for a day.

Dry Ginger

The Ginger is harvested after eight months for producing dry Ginger. The harvested Ginger should be soaked in water for 6–7 hours. Then, the rhizomes should be rubbed well to clean the extraneous matter. After cleaning, the rhizomes should be removed from water and the outer skin should be removed with bamboo splinters having pointed ends. Deep scraping may be avoided to prevent damage of oil cells which are just below the outer skin. The peeled rhizomes should be washed and dried uniformly in Sun for one week. The dry rhizomes should be rubbed together to get rid of the last bit of skin or dirt. The yield of dry Ginger is 19–25 per cent of fresh Ginger depending on the variety and location where the crop is grown. Fresh Ginger (with relatively low fibre) harvested at 170–180 DAP (Days after planting) can be used for preparing salted Ginger.

Tender Ginger

The rhizomes with a portion of the pseudostem may be washed thoroughly and soaked in 30 per cent salt solution containing 1 per cent citric acid. After 14 d, it is ready for use and can be stored under refrigeration.

3.12. STORAGE OF SEED RHIZOME

The seed rhizomes should be stored properly in pits under shade. For seed material, bold and healthy rhizomes from disease–free plants should be selected immediately after the harvest. For this purpose, healthy and disease–free clumps should be marked in the field when the crop is 6–8 month old and still green. The seed rhizomes should be treated with a solution containing Quinalphos 30 EC 0.075 per cent and Mancozeb 50 WP 0.3 per cent for 30 min and dried under shade. The seed rhizomes are stored in pits of convenient size in sheds.

3.13. USES

- Ginger, an herb, is used as a spice. Whole fresh Ginger root provides the fresh taste. It is used in vegetable curries, tea, marinades etc. and for baking. Fresh Ginger can be kept for several weeks. Dry Ginger (Hindi name: Saunth) is also used in many Indian dishes.

- Ginger is prized for its flavour and medicinal properties. In addition to its use as spice and condiment, it is also used to treat liver complaints, flatulence, anemia, rheumatism, piles and jaundice in Indian systems of medicines (Aired, Sridhar and Unani).
- Ginger root (Biological name: Rhizome) contains cancer–fighting antioxidants, hence useful for fighting factors that cause blood clots.
- The dried root contains approximately 4 per cent volatile oils that give Ginger its characteristic odour and taste. Fresh Ginger, dry Ginger powder, oleoresin and oil are used in food processing. It is indispensable in the manufacture of Ginger bread, confectionary, Ginger ale, curry powders, certain curried meats, table sauces, in pickling and in the manufacture of certain cordials, Ginger cocktail, carbonate drinks, liquors etc.
- In medicine, it is used as carminative and stimulant. It has wider applications in indigenous medicines. Ginger oil is used as food flavourant in soft drinks.

3.14. PLANT PROTECTION

Insect pests

1. Shoot borer

Scientific name - *Conogethes punctiferalis*

Shoot borer is most destructive insect pest of Ginger.

Nature of damage

The larvae feed on the leaves after emerging from egg and bore the shoots and feed inside. As a result, leaves of infested pseudo stems turn yellow in colour and dry–off later.

Control measures

The insect can be controlled by spraying any one of the insecticides given in Table 35.

Table 35. Insecticides used for controlling the Shoot borer of Ginger

Sr. No.	Name of insecticide	Amount (per litre of water)
1.	Acephate 75 WP	0.75 g
2.	Thiomethoxam 25 WG	0.2 g
3.	Carbosulphan 25 EC	2.0 ml
4.	Imidachloropid 17.8 SL	0.2 ml
5.	Acetamiprid 20 SL	0.1 g

Larva

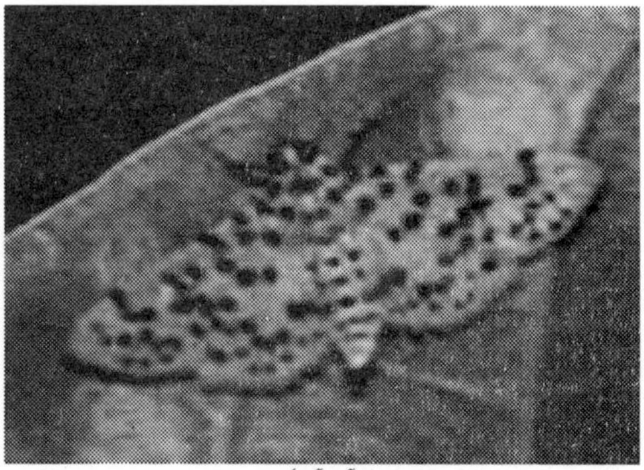

Adult

Fig. 10. Shoot borer of Ginger

2. Leaf roller

Scientific name - *Udaspes folus*

Leaf roller causes damage to Ginger by rolling the leaf, which curtails the photosynthetic area of the leaf.

Larva

Adult

Fig. 11. Ginger leaf roller

Nature of damage

The larvae of insect pest attack on leaves. Leaves become folded or rolled longitudinally. In severe cases, this insect pest causes complete defoliation.

Control measures

- The clipping of folding leaves and destroying by burning or burrowing under soil is an effective method of controlling this insect pest in initial stage.

- The insect pest can be controlled by spraying any one of the insecticides given in Table 36.

Table 36. Insecticides used for controlling the Leaf roller of Ginger

Sr. No.	Name of insecticide	Amount (per litre of water)
1.	Acephate 75 WP	0.75 g
2.	Thiomethoxam 25 WG	0.2 g
3.	Cartap Hydrochloride 50 SP	1.0 g
4.	Spinosad 45 SC	0.25 ml

3. Rhizome scale

Scientific name - *Aspidiella hartii*

Rhizome scale insect causes damage both in the field and in storage.

Nature of Damage

- **Field infestation** - Rhizome scale insect sucks the sap from the affected parts. The insect–affected plants look devitalized, pale and withered before drying completely.
- **Storage infestation** - In initial stage of infestation, the white–coloured scales are seen scattered on rhizomes and latter they congregate near the growing buds. When the infestation is severe, the rhizome and buds shrivel and ultimately the entire rhizome dries.

Control measures

- The insect–affected plants should be collected and destroyed by burning or burrowing under the soil.
- The seed materials collected from the field that are free from scale infestation should be used for planting. The seed material should be treated with Quinalphos 20 EC (@0.075 per cent for 20–30 minutes before storage and also before sowing in case the infestation persists.

4. Thrip

Scientific name - *Panchaetothrips indicus*

Nature of damage

The adult and nymphs suck the sap and cause damage to leaves. The insect–affected leaves turn pale, rolled up and gradually dry up.

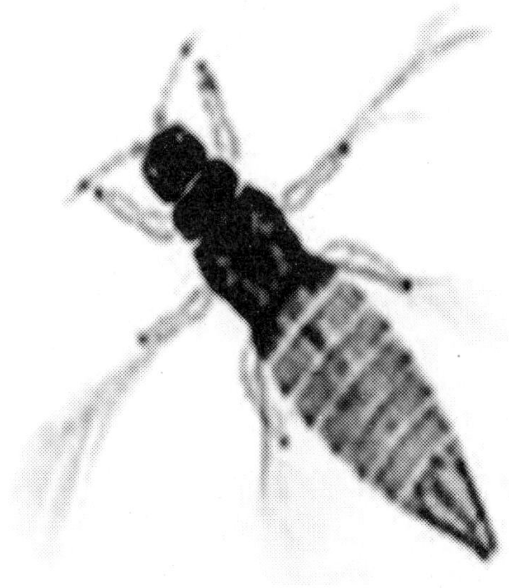

Fig. 12. Thrips (Adult)

Control measures

- The dry drooping leaves, dry leaf sheaths and other dry plant parts should be destroyed by burning.
- The thrips can be controlled by spraying any one of the insecticides in the Table 37.

Table 37. Insecticides used for controlling the Thrips of Ginger

Sr. No.	Name of insecticide	Amount (per litre of water)
1.	Acephate 75 WP	0.75 g
2.	Quinalphos 25 EC	2.0 ml
3.	Clothianidin 50 WDG	0.1 gm
4.	Triazophos 40 EC	2.0 ml

Diseases

1. Soft rot or rhizome rot

C.O.: *Pythium aphanidermatum*, *Pythium vexans* and, *Pythium myriotylum*

Soft rot is the most destructive disease of Ginger, which results in total loss of affected clumps. Younger sprouts are the most susceptible to this pathogen. The fungus multiplies with build–up of soil moisture with the onset of southwest monsoon.

Symptoms

Foliar symptoms appear as light yellowing of the tips of lower leaves, which gradually spreads to the leaf blades. In early stages of the disease, the middle portion of the leaves remain green while the margins become yellow. The yellowing spreads to all leaves of the plant from the lower region to the upward region and is followed by drooping, withering and drying of pseudostems.

Younger sprouts are the most susceptible to the pathogen. The infection starts at the collar region of the pseudostem and progresses upwards as well as downwards. The collar region of the affected pseudostem becomes water–soaked and the rotting spreads to the rhizome resulting in soft rot. At a later stage root infection is also noticed (Srinivasan et al 2008)

Control measures

- Seed rhizome should be collected from the field free from disease as the disease is seed borne.
- The seed rhizome should be treated with Mancozeb 50 WP @ 0.3 per cent for 30 min before storage and once again before planting. This reduces the incidence of the disease.
- Cultural practices such as selection of well–drained soils for planting is important for managing the disease, since stagnation of water predisposes the plant to infection.
- Application of *Trichoderma harzianum* along with neem cake @ 1 kg per bed helps in preventing the disease.
- Once the disease is located in the field, removal of affected clumps and drenching the affected and surrounding beds with any one of the fungicides given in Table 38 is found effective in controlling the disease.

Table 38. Fungicides used for controlling the Soft rot or rhizome rot of Ginger

Sr. No.	Name of fungicide	Amount (per litre of water)
1.	Copper oxychloride 50 WP	5.0 g
2.	Carbendazim 50 WP	1.0 g
3.	Mancozeb 75 WP	2.5 g
4.	Chlorothalonil 75 WP	2.0 g

2. Leaf spot

C.O.: *Phyllosticta zingiberi*

Leaf spot is a disease of leaf and the disease is noticed on the leaves from July to October. The incidence of the disease is severe in Ginger grown under exposed conditions.

Symptoms

The disease starts as a water–soaked spot and later it turns as a white spot surrounded by dark brown margins and yellow halo. The lesions enlarge and adjacent lesions coalesce to form necrotic areas.

Control measures

The disease can be controlled by spraying any one of the fungicides given in Table 39.

Table 39. Fungicides used for controlling Leaf spot of Ginger

Sr. No.	Name of fungicide	Amount (per litre of water)
1.	Copper oxychloride 50 WP	5.0 g
2.	Carbendazim 50 WP	1.0 g
3.	Mancozeb 75 WP	2.5 g
4.	Chlorothalonil 75 WP	2.0 g

3. Bacterial wilt

C.O.: *Pseudomonas solanacearum*

Bacterial wilt of Ginger is a very destructive, parasitic disease and has been reported from most areas of Hawaii where Ginger is grown commercially.

Symptoms

The initial symptoms of wilt are a slight yellowing and wilting of the lower leaves, which gradually progresses to the upper leaves. In the advanced stage, the plants exhibit severe yellowing and wilting symptoms. The wilt progresses upwards, affecting the younger leaves, followed by a complete yellowing and browning of the entire shoot. Under conditions favourable for disease development, the entire shoot becomes flaccid and wilts with little or no visible yellowing. However, the plant dries very rapidly and the foliage becomes yellow–brown in 3–4 d. Water–soaked spots appear at the collar region of the pseudostem and progresses upwards and downwards. In advanced infections, the entire rhizome becomes soft and rots. The affected pseudostem and rhizome when pressed gently extrudes milky ooze from the vascular strands.

Control measures

- The cultural practices adopted for managing soft rot should also be adopted for bacterial wilt.

- The seed rhizomes may be treated with Streptocycline 200 ppm for 30 min and shade–dried before planting.
- Once the disease is noticed in the field, all beds should be drenched with Bordeaux mixture @ 1% or Copper oxychloride 50 WP @ 0.2 per cent.

4 TURMERIC

Scientific name - *Curcuma longa*

Family - Zingiberaceae

Commercial part - Rhizome

Indian names - *Bengali* : Halud; *English* : Turmeric; *Hindi* : Haldi; *Assamese* : Halodhi; *Gujarati* : Halad or Haldar; *Malayalam*: Manjal; *Marathi* : Halad; *Punjabi* : Haldi; *Tamil* : Manjal; *Telugu* : Haridra.

Turmeric is a very important spice in India. The name Turmeric appears to derive from the Latin, *terra merita* (merited Earth) or turmeryte, possibly related to Saffron. The tuberous rhizomes or underground stems of Turmeric (*Curcuma longa*) is used as condiment, dye, drug and cosmetic in addition to its use in religious ceremonies. India is a leading producer and exporter of Turmeric in the world. Andhra Pradesh, Tamil Nadu, Orissa, Karnataka, West Bengal, Gujarat, Meghalaya, Maharashtra, Assam are some of the important States that cultivate turmeric, of which, Andhra Pradesh alone occupies 35.0 per cent of area and 47.0 per cent of production. Turmeric (*Curcuma longa*) is native to Asia and India. Turmeric is also cultivated in China, Myanmar, Nigeria and Bangladesh. However, authentic figures about area and production are not available. Major area is in India, which constitutes 82 per cent followed by China (8 per cent), Myanmar (4 per cent), Nigeria (3 per cent) and Bangladesh (3 per cent).

The main turmeric–producing States in India are Andhra Pradesh, Tamil Nadu, Orissa, Karnataka, West- Bengal, Gujarat and Kerala. Turmeric is traditionally used in India for medicinal, religious, culinary purposes and also as a cosmetic and dye (Shah 1997). In Ayurveda, Turmeric is regarded as aromatic stimulant, tonic, carminative and antihelminthic. The essential oil of Turmeric is antiseptic and it is used in treating gallstones and gall complaints (Pruthi 1976).

4.1. BOTANICAL DESCRIPTION

Turmeric (*Curcuma longa*) is a rhizomatous, herbaceous, perennial plant of the Ginger family, zingiberaceae. Turmeric is the boiled, dried, cleaned and polished rhizomes of *Curcuma longa*.

- **Plant** - It is an herbaceous, perennial, 60–90 cm tall plant with short stem and tufted leaf, and with a thick underground rhizome giving rise to primary and secondary rhizome called 'fingers'.
- **Leaf** - The number of leaves is 7–12 and is broadly lanceolate with long leaf stalk. The leaf sheaths form the pseudostem. The lamina is green above and pale green below and has a length of 30–40 cm and width 8–12 cm.
- **Flower** - Flowers are borne on separate peduncle arising directly from the rhizome.
- **Inflorescence** - A central spike is of 10–15 cm length. 1–4 flowers are borne in axil of the bract opening one at a time. About 30 flowers are produced in a spike.
- **Seeds** - Seeds are produced in capsules and there may be one to numerous sunken capsules in an inflorescence.

4.2. ORIGIN AND DISTRIBUTION

Turmeric is a native of India. Apart from India, it is cultivated in Pakistan, Malaysia, Myanmar, Vietnam, Thailand, Philippines, Japan, Korea, China, Sri Lanka, Nepal, East and West Africa, South Pacific Islands, Malagasy, Caribbean Islands and Central America. In India, it is cultivated in the States of Andhra Pradesh, Maharashtra, Orissa, Tamil Nadu, Karnataka and Kerala.

4.3. CLIMATE

Turmeric is a tropical crop requiring warm and humid climate and cultivated from sea level to 1200 m MSL. Turmeric needs temperatures ranging between 20°C–35°C with annual rainfall of 1,500 mm or more, under rainfed or irrigated conditions. Growth ceases when temperature falls below 20°C, and hence early–planted Turmeric gives good yield.

4.4. SOIL

Turmeric can be grown on different types of soil such as sandy loam to clay loam or alluvial soils. Soil should be rich in organic matter and uniform in texture. Rich loamy soils having natural drainage and irrigation facilities are the best for turmeric. Turmeric cannot withstand water stagnation or alkalinity. But it thrives best in well–drained sandy loam soil. It can also grow in light black,

black clayey loams and red soils in irrigated and rainfed conditions. The soil pH range of 4.5–7.5 is ideal for Turmeric cultivation.

Fig. 13. Turmeric

4.5. LAND PREPARATION

The land should be prepared with the receipt of early monsoon showers. The soil should be brought to a fine tilth by giving deep ploughing followed by planking. Hydrated lime @ 500 kg/ha should be applied for laterite soils and it should be thoroughly mixed by ploughing followed by planking. Immediately with the receipt of pre–monsoon showers, beds of 1.0 m width, 15 cm height and of convenient length should be prepared with spacing of 50 cm between beds. Planting can also be done by forming ridges and furrows.

4.6. PLANTING

1. Seed materials

Rhizomes are used as planting material. Whole or split mother and finger rhizomes should be used for planting. Well–developed, healthy and disease–free rhizomes weighing about 35–45 g should be used for planting. The following points need to be remembered before planting turmeric:

- It is necessary to store the seed rhizomes for 2–3 months from harvesting to planting. This may be done by spreading them thinly under a cover of Turmeric leaves or storing them by treating the rhizomes with Mancozeb 50 WP @ 2.5 g per litre of water and Malathion 50 EC @ 2 ml per litre for 30 min before storing heaps under a layer of straw and soil.
- Well–developed, healthy and disease–free rhizomes should be selected.
- Whole or split mother rhizomes weighing 30 to 35 g should be used for planting.

2. Seed treatment

The seed materials should be treated with any one of the fungicides/bio–fungicides in the Table 40 before planting.

Table 40. Fungicides / bio-fungicides used for the seed treatment of Turmeric

Sr. No.	Name of fungicide/bio–fungicide	Amount (per litre of water)
1.	Carbendazim 50 WP	2.0–3.0 g
2.	M.E.M.C 6 WP	1.0 g
3.	Captan 50 WP	2.0–5.0 g
4.	Mancozeb 75 WP	2.5 g
5.	*Trichoderma viridi*	4.0–5.0 g

The treated seeds should be kept under shade for at least 24 hours and then planted.

3. Season

Plants should be gathered annually for their rhizomes, and propagated from some of those rhizomes in the following season.

- Planting season varies with the area of cultivation and variety.
- Planting should be done during May–June or July–August in different tracts.
- In Kerala and other west coast areas where rainfall is sufficiently early, the crop can be planted during April–May with the receipt of pre–monsoon showers.
- In Andhra Pradesh, planting for short duration varieties such as Kasturi is done during second fortnight of May, and for medium duration varieties such as Kesari, planting is done during 1st fortnight of June.
- For long duration varieties such as Mydukur, planting is done between 2nd fortnight of June to 2nd fortnight of July.

4. Method of planting

1. Furrow and ridge method

Planting is done by dibbling rhizome in furrows made by the country plough. The seeds are covered with loose soil from the ridge.

In Andhra Pradesh, the broad ridge method of planting is superior and more profitable than the ridge and furrow method, as the elevated beds provide better drainage.

The crop can be planted on flat beds or on ridges, or large scale potato planter can be used for Turmeric planting.

Emergence of seedlings takes place 2–4 weeks after sowing.

2. Pit method

Small pits are made with a hand hoe on the beds with a spacing of 25 x 30 cm. Pits are filled with well– decomposed cattle manure or compost, seed rhizomes are placed over it and it is covered with soil. The optimum spacing in furrows and ridges is 45–60 cm between the rows and 25 cm between the plants.

3. Mixed cropping

Turmeric can be grown as an inter–crop with chillies, colocasia, onion, brinjal and cereals such as maize, ragi etc.

Spacing

Black heavy soils - 45 x 22.5 cm

Red loamy soils - 30 x 15 cm

Over all, the optimum spacing should be 45–60 cm between rows and 22.5 cm within the row. In case of planting in beds, rhizomes are planted 25–35 cm apart in each direction.

Seed rate

Mother Rhizomes - 2,000–2,500 kg/ha as a sole crop

Primary fingers -1,500–2,000 kg/ha as a sole crop

Fruit gardens - 400–500 kg/ha as an inter-crop

Top dressing should be completed within 120 days after sowing. After fertilization, the field is given a light hoeing and the clumps are earthed up and irrigated.

4.7. MANURING

Manures

Compost or FYM: 25.0–30.0 t/ha

Fertilizers

Nitrogen (N)	: 60.0 kg/ha
Phosphorus (P_2O_5)	: 50.0 kg/ha
Potassium (K_2O)	: 120.0 kg/ha
Zinc (Zn)	: 5.0 kg/ha

Method of application

Manures

The manures to be applied to the soil should be broadcasted during land preparation and ploughed at the time of land preparation or should be applied as basal dressing by spreading over the beds or into the pits at the time of planting. Organic manures such as oil organic cakes can also be applied @ 2 t/ha. In such case, the dosage of FYM can be reduced. Integrated application of coir compost (combined with FYM, biofertilizer (*Azospirillum*) and half recommended dose of NPK is integrated application of coir compost @ 2.5 t/ha) is also recommended.

Fertilizers

Basal dressing: 50.0 kg P_2O_5/ha

Top dressing

- 45 DAP - 30.0 kg N and 60.0 kg K_2O/ha
- 90 DAP - 30.0 kg N and 60.0 kg K_2O/ha

Top dressing should be completed within 120 days after sowing. After fertilization, the field is given a light hoeing and the clumps are earthed up and irrigated.

Micronutrients

The following amount of micronutrient, dissolved in 250 litres of water, should be sprayed on the crop to correct the deficiency of micronutrients:

Ferrous sulphate - 375 g

Zinc sulphate - 375 g

Borax - 375 g

4.8. INTERCULTURAL OPERATION

1. Weeding

Weeds pose a problem in the production of turmeric. So land should be kept free from weeds. Weeding should be done thrice at 60, 90 and 120 days after planting depending upon weed intensity.

2. Mulching

The crop should be mulched with green leaves @ 12-15 t/ha immediately after planting. Mulching may be repeated @ 7.5 t/ha at 45 and 90 days after planting, after weeding, application of fertilizers, and earthing up.

4.9. IRRIGATION

Irrigation should be given according to the need of the crop. The first and second irrigation may be given either before or after planting of the crop and 3–4 days after planting of the crop respectively. After that it should be given at 8–10 d interval. But water stagnation should be avoided during rainy season. In the case of irrigated crop, depending upon the weather and soil conditions, about 15–23 irrigations should be given in clayey soils and 40 irrigations in sandy loam soils.

Table 41. Some of the commercial and improved varieties

S.N.	Variety	Mean yield (fresh) t/ha	Crop duration	Dry recovery (%)	Curcumin (%)	Oleoresin (%)	Essential oil (%)
1.	Suvarna	17.4	200	4.3	13.5	4.5	7.0
2.	Suguna	29.3	190	12.0	7.3	13.3	6.0
3.	Sudarsana	28.8	190	12.0	5.3	15.0	7.0
4.	IISR Prabha	37.5	195	19.5	6.5	15.0	6.5
5.	IISR Prathibha	39.1	188	18.5	6.2	16.2	6.2
6.	CO-1	30.0	285	19.5	3.2	6.7	3.2
7.	BSR-112	30.7	285	20.5	4.2	4.0	3.7
8.	Krishna	9.2	240	16.4	2.8	3.8	2.0
9.	Sugandham	15.0	210	23.5	3.1	11.0	2.7
10.	Roma	20.7	250	31.0	9.3	13.2	4.2
11.	Suroma	20.0	255	26.0	9.3	13.1	4.4
12.	Ranga	29.9	250	24.0	6.3	13.5	4.4
13.	Rasmi	31.3	240	23.0	6.4	13.0	4.4
14.	Rajendra Sonia	42.0	225	18.0	8.4	15.0	5.0
15.	Supreme	35.4	210	19.3	6.0	16.0	4.0
16.	IISR Kedaram	34.5	210	18.9	5.5	13.6	3.0

4.10. MIXED CROPPING

Turmeric can be grown as an inter–crop with crops such as chilies, colocasia, onion, brinjal and cereals such as maize, ragi etc.

4.11. HARVESTING

Depending upon the variety, the crop becomes ready for harvest in 7–9 months after planting during January–March.

Early varieties: 7–8 months

Medium varieties: 8–9 months

Late varieties: 9 months

The land is ploughed and the rhizomes are gathered by handpicking. The clumps are carefully lifted with a spade. The harvested rhizomes are cleared of mud and other extraneous matter adhering to them.

4.12. YIELD

20.0–25.0 q/ha

4.13. PLANT PROTECTION

Insect pests

1. Shoot borer

Scientific name - *Conogethes punctiferalis*

Shoot borer is most destructive insect pest of Turmeric.

Nature of damage

The larva feeds on the leaves after emerging from the egg, bores the shoots and feeds inside. As a result, leaves of infested pseudostems turn yellow in colour and dry afterwards.

Control measures

The insect can be controlled by spraying any one of the insecticides in the Table 42.

Table 42. Insecticides used for controlling the Shoot borer of Turmeric

Sr. No.	Name of insecticide	Amount (per litre of water)
1.	Acephate 75 WP	0.75 g
2.	Thiomethoxam 25 WG	0.2 g
3.	Carbosulphan 25 EC	2.0 ml
4.	Imidachloropid 17.8 SL	0.2 ml
5.	Acetamiprid 20 SL	0.1 g

2. Leaf roller

Scientific name - *Udaspes folus*

Leaf roller causes damage to Turmeric by rolling the leaf, which curtails the photosynthetic area of the leaf.

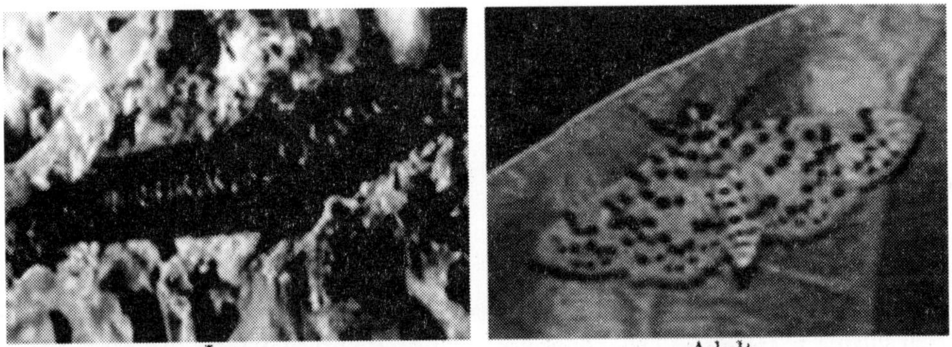

Larva Adult

Fig 14. Shoot borer of Ginger

Nature of damage

The larvae of insect pest attack on the leaves and the leaves become folded or rolled longitudinally. In severe cases, this insect pest causes loss in yield.

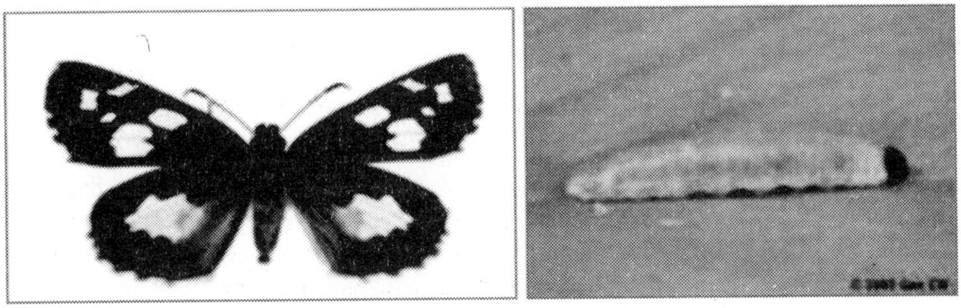

Larva Adult

Fig 15. Ginger leaf roller

Control measures

- The clipping of folding leaves and destroying by burning or burrowing under soil is an effective method of controlling this insect pest in initial stages.
- The insect pest can be controlled by spraying any one of the insecticides given in Table 43.

Table 43. Insecticides used for controlling the Leaf roller of Turmeric

Sr. No.	Name of insecticide	Amount (per litre of water)
1.	Acephate 75 WP	0.75 g
2.	Thiomethoxam 25 WG	0.2 g
3.	Cartap Hydrochloride 50 SP	1.0 g
4.	Spinosad 45 SC	0.25 ml

3. Rhizome scale insect

Scientific name - *Aspidiella hartii*

Rizome scale insect causes damage both in the field and in storage.

Nature of damage

Field infestation

Rhizome scale insect sucks the sap from the affected parts. The insect–affected plants look devitalized, pale and withered before drying completely.

Storage infestation

In initial stage of infestation, the white–coloured scales are seen scattered on rhizomes and later they congregate near the growing buds. When the infestation is severe, the rhizome and buds shrivel and ultimately the entire rhizome dries off.

Control measures

- The insect–affected plants should be collected and destroyed by burning or burrowing under the soil.
- The seed materials collected from the field that are free from scale infestation should be used for planting. The seed material should be treated with Quinalphos 20 EC (@ 0.075% for 20–30 min before storage and also before sowing in case the infestation persists.

4. Thrips

Scientific name - *Panchaetothrips indicus*

Nature of damage

The adult and nymphs suck the sap and cause damage to the leaves. The insect–affected leaves turn pale, rolled up and gradually dry up.

Control measures

- The dry, drooping leaves, dry leaf sheaths and other dry plant parts should be destroyed by burning.
- The thrips can be controlled by spraying any one of the insecticides given in Table 44.

Table 44. Insecticides used for controlling the Thrips of Turmeric

Sr. No.	Name of insecticide	Amount (per litre of water)
1.	Acephate 75 WP	0.75 g
2.	Quinalphos 25 EC	2.0 ml
3.	Clothianidin 50 WDG	0.1 gm
4.	Triazophos 40 EC	2.0 ml

Diseases

1. Soft rot or rhizome rot

C.O.: *Pythium aphanidermatum*, *Pythium vexans* and *Pythium myriotylum*

Soft rot is the most destructive disease in *turmeric*, which results in total loss of affected clumps. Younger sprouts are the most susceptible to this pathogen.

The fungus multiplies with build–up of soil moisture with the onset of south west monsoon.

Symptoms

Foliar symptoms appear as light yellowing of the tips of lower leaves, which gradually spread to the leaf blades. In early stages of the disease, the middle portion of the leaves remains green while the margins become yellow. The yellowing spreads to all leaves of the plant from the lower region to the upwards region and is followed by drooping, withering and drying of pseudostems.

Younger sprouts are the most susceptible to this pathogen. The infection starts at the collar region of the pseudostem and progresses upwards as well as downwards. The collar region of the affected pseudostem becomes water–soaked and the rotting spreads to the rhizome resulting in soft rot. At a later stage, root infection is also noticed. (Srinivasan et al 2008)

Control measures

- Seed rhizome should be collected from the field free from disease as the disease is seed borne.
- The seed rhizome should be treated with Mancozeb 50 WP @ 0.3 per cent for 30 min before storage and once before planting. This reduces the incidence of the disease.
- Cultural practices such as selection of well–drained soils for planting is important for managing the disease, since stagnation of water pre–disposes the plant to infection.

- Application of *Trichoderma harzianum* along with neem cake @ 1 kg/bed helps in preventing the disease.
- Once the disease is located in the field, removal of affected clumps and drenching the affected and surrounding beds with any one of the fungicides given in Table 45 is found effective in controlling the disease:

Table 45. Fungicides used for controlling the Soft rot of rhizome rot of turmeric

Sr. No.	Name of fungicide	Amount (per litre of water)
1.	Coper oxychloride 50 WP	5.0 g
2.	Carbendazim 50 WP	1.0 g
3.	Mancozeb75 WP	2.5 g
4.	Chlorothalonil 75 WP	2.0 g

2. Leaf spot

C.O.: *Phyllosticta zingiberi*

Leaf spot is a disease of leaf and the disease is noticed on the leaves from July–October. The incidence of the disease is severe in *Turmeric* grown under exposed conditions.

The disease can be controlled by spraying any one of the fungicides given in Table 46.

Table 46. Fungicides used for controlling the Leaf spot or Turmeric

Sr. No.	Name of fungicide	Amount (per litre of water)
1.	Coper oxychloride 50 WP	5.0 g
2.	Carbendazim 50 WP	1.0 g
3.	Mancozeb 75 WP	2.5 g
4.	Chlorothalonil 75 WP	2.0 g

Symptoms

The disease starts as a water–soaked spot and later turns as a white spot surrounded by dark brown margins and yellow halo. The lesions enlarge and adjacent lesions coalesce to form necrotic areas.

3. Bacterial wilt

C.O.: *Pseudomonas solanacearum*

Bacterial wilt of *Turmeric* is a very destructive, parasitic disease and has been reported from most areas of Hawaii where Turmeric is grown commercially.

Symptoms

The initial symptoms of wilt are slight yellowing and wilting of the lower leaves, which gradually progresses to the upper leaves. In the advanced stage, the plants exhibit severe yellowing and wilting symptoms. The wilt progresses upward, affecting the younger leaves, followed by a complete yellowing and browning of the entire shoot. Under conditions favourable for disease development, the entire shoot becomes flaccid and wilts with little or no visible yellowing. However, the plant dries very rapidly and the foliage becomes yellow–brown in 3–4 days. Water–soaked spots appear at the collar region of the pseudostem and progresses upwards and downwards. In advanced infections, the entire rhizome becomes soft and rots. The affected pseudostem and rhizome when pressed gently extrudes a milky ooze from the vascular strands.

Control measures

The cultural practices adopted for managing soft rot should also be adopted for bacterial wilt.

- The seed rhizomes may be treated with Streptocycline 200 ppm for 30 min and shade–dried before planting.
- Once the disease is noticed in the field all beds should be drenched with Bordeaux mixture @ 1% or Copper oxychloride 50 WP @ 0.2%.

4.14. PROCESSING

1. Curing

Fresh Turmeric is cured for obtaining dry turmeric. Curing involves boiling of fresh rhizomes in water and drying in the Sun. The curing is done by two methods:

- **Traditional method**: In the traditional method of curing, the cleaned rhizomes are boiled in water, just enough to immerse them, for 45–60 min. Boiling is stopped when froth comes out and white fumes appear giving out a typical odour. Over–cooking spoils the colour of the final product while under–cooking renders the dried product brittle.
- **Improved method:** In the improved scientific method of curing, the cleaned fingers (approximately 50 kg) are taken in a perforated trough of 0.9 x 0.5 x 0.4 m size made of GI or MS sheet with extended parallel handle. Water is poured into the trough so as to immerse the Turmeric fingers. The whole mass is boiled till the fingers become soft. The soft fingers are taken out from the trough and dried in the Sun.

2. Drying

The cooked fingers are dried in the Sun by spreading them in 5–7 cm thick layers on bamboo mats or are dried through floor aeration. It may take 10–15 d for the rhizomes to become completely dry.

3. Polishing

Dried Turmeric has a poor appearance and a rough dull outer surface with scales and root bits. The appearance is improved by smoothening and polishing the outer surface by manual or mechanical rubbing.

4. Storing

Rhizomes for seed purpose are generally stored by heaping in well–ventilated rooms and covered with Turmeric leaves.

4.15. USES

There are 20 reasons to add Turmeric to your diet as follows:

1. It is a natural antiseptic and antibacterial agent, useful in disinfecting cuts and burns.
2. When combined with cauliflower, it has shown to prevent prostate cancer and stop the growth of existing prostate cancer.
3. It prevents breast cancer from spreading to the lungs in mice.
4. It may prevent melanoma and causes existing melanoma cells to commit suicide.
5. It reduces the risk of childhood leukaemia.
6. It is a natural liver detoxifier.
7. It may prevent and slow down the progression of Alzheimer's disease by removing amyloyd plaque built–up in the brain.
8. It may prevent metastases from occurring in many different forms of cancer.
9. It is a potent natural anti-inflammatory that works as well as many anti-inflammatory drugs but without side effects.
10. It has shown promise in slowing the progression of multiple sclerosis in mice.
11. It is a natural painkiller and cox-2 inhibitor.
12. It may aid in fat metabolism and help in weight management.

Source: http://www.healthdiaries.com/eatthis/20-health-benefits-of-turmeric.html

5 CORIANDER

Scientific name - *Coriandrum sativum*

Family - *Apiaceae*

Commercial part - Seed

Indian names - *Bengali* : Dhana, Dhania; *Hindi* : Dhania or Dhanya; *Gujarati* : Kothmiri, Libdhaba; *Kannada* : Kothambri, Kothamiri bija; *Kashmiri* : Deaniwal, Kothambalari; *Malayalam* : Kothumpalari bija; *Marathi* : Dhana; *Oriya* : Dhania; *Punjabi* : Dhania; *Sanskrit* : Dhanyaka; *Tamil* : Kothamalli; *Telugu* : Dhaniyal

Coriander is an annual herb in the family *Apiaceae.* It is also known as dhania or Chinese parsley or cilantro, particularly in America. Coriander is an important spice crop having a prime position in flavouring food. The name Coriandrum, used by Pliny, is derived from *koros,* (a bug), in reference to the foetid smell of the leaves. As an aromatic stimulant and spice, it has been cultivated and used from very ancient times. It was employed by Hippocrates and other Greek physicians.

Table 47. Nutritional value of Coriander leaves per 100 g

Energy	95 kJ (23 kcal)
Carbohydrates	4 g
Dietary fibre	3 g
Fat	0.5 g
Protein	2 g
Vitamin A equiv.	337 μg (42%)
Vitamin C	27 mg (33%)

5.1. ORIGIN AND DISTRIBUTION

Coriander is native to regions spanning from southern Europe and North Africa to south-western Asia and commercially produced in India, Morocco, Russia, East European countries, France, Central America, Mexico and USA.

It is cultivated in Rajasthan, Gujarat, Madhya Pradesh, Tamil Nadu, Uttar Pradesh etc. It is mainly used as a condiment for its medicinal properties as well as for culinary purposes as spice. Green leaves of Coriander are also extensively used for culinary purposes.

Fig. 16. Coriander plant and seed

5.2. BOTANICAL DESCRIPTION

- **Plant** - It is a thin stemmed, small, bushy herb, having 25–50 cm height with many branches and umbels.
- **Leaf** - The leaves are alternate and compound. The leaves are variable in shape, broadly lobbed at the base of the plant, and slender and feathery on the higher flowering stems.
- **Inflorescence** - The inflorescence is compound umbel comprising of five smaller umbels.
- **Fruit** - The fruit has a delicate fragrance. Seeds are pale white to light brown in colour and globular in shape with 3–4 mm diameter. Seeds, when pressed, break into two locules each having one seed. The dry fruits are known as Coriander or *coriandi* seeds. There are two distinct morphological types, one erect and tall with a comparatively stronger main shoot and the other bushy with a relatively weaker main shoot and longer spreading branches.

5.3. CLIMATE

Coriander is a tropical crop and can be successfully cultivated as a rabi season crop in an area free from severe frost at the time of flowering and seed formation. Cool weather is favourable for its cultivation. It is susceptible to mild frost. Coriander requires cool climate during growth stage and warm dry climate at maturity. It is observed that quality of seed is superior and essential oil content is more when the crop is grown in colder regions and at high altitudes. Although the crop is heat–loving, it has an optimum growing temperature of 18°C. Being a tropical crop, Coriander plants prefer frost–free tropical climate.

5.4. SOIL

Coriander grows well in almost all types of soils. But sandy loam and clay soils are best suited for the cultivation of Coriander. Coriander can be successfully grown in black soils also where the crop is raised as a rainfed crop. Coriander is not tolerant to alkaline and saline soils. Suitable pH ranges are from 4.5–8.0, with an optimum of 6.3.

5.5. LAND PREPARATION

Land should be ploughed 2–3 times followed by planking, to bring the soil to fine tilth. Clods should be broken and stubbles of previous crop should be removed. Before land preparation, pre–sowing irrigation should be given, if optimum moisture for seed germination is not available in the soil.

5.6. SOWING OF SEED

1. Time of sowing

Usually Coriander is cultivated during Rabi season. The best time for sowing of Coriander is 15th October to 15th November. Timely sowing is beneficial as in early sowing germination is affected on account of excess sunshine and late sowing leads to paltry growth of plants and development of various diseases. Sowing of crop at a fixed interval should be carried out for regular marketing of the crop.

2. Seed rate

15.0–20.0 kg/ha for production of seed and 40.0 kg/ha for production of green leaves.

3. Seed treatment

The seed materials should be treated with any one of the fungicides/bio–fungicides given in the Table 48 before planting.

The treated seeds should be kept under shade for at least 24 hours and then planted.

Table 48. Fungicides/ bio-fungicides used for seed treatment of Coriander

Sr. No.	Name of fungicide/bio–fungicide	Amount (per litre of water)
1.	Carbendazim 50 WP	2.0–3.0 g
2.	M.E.M.C 6 WP	1.0 g
3.	Captan 50 WP	2.0–5.0 g
4.	Mancozeb 75 WP	2.5 g
5.	Trichoderma viridi	4.0–5.0 g

4. Method of sowing

Seed should be rubbed to split into two halves before sowing. While rubbing, care should be taken to save the sprouting portion. Sowing is generally done by broadcasting method but sowing may also be done in rows at spacing of 30 cm apart. Seeds should be soaked in water overnight to remove inhibitor and thereby enhance germination.

5.7. MANURING

1. Rainfed condition

Manures

Compost or FYM: 15.0–20.0 t/ha

Fertilizers

Nitrogen (N) : 20.0 kg/ha
Phosphorus (P_2O_5) : 30.0 kg/ha
Potassium (K_2O) : 20.0 kg/ha

2. Irrigated condition

Manures

FYM or Compost: 20.0 t/ha

Fertilizers

Nitogen (N) : 60.0 kg/ha
Phosphorus (P_2O_5) : 30.0 kg/ha
Potassium (K_2O) : 20 kg/ha

- **Method of application**

Manures

The total amount of manures should be applied at the time of land preparation and should be mixed while ploughing followed by planking.

Fertilizers

One third dose of nitrogen (20 kg), full dose of phosphorous and potash should be applied as basal dose and remaining 40 kg nitrogen should be top dressed in two equal split doses at the time of first irrigation and at flowering stage.

5.8. INTERCULTURAL OPERATION

1. Weeding

Weeds pose a problem in the cultivation of Coriander. During early stages of growth, Coriander plants grow very slowly. So weeding during this period is very essential to save the plants from weed competition. The weeding schedule for rainfed and irrigated crops is as follows:

- Rainfed crop - 25–30 DAS (Days After Sowing)
- Irrigated crop - (i) First Weeding: 40–45 DAS (ii) Second weeding: 50–60 DAS

2. Thinning

Thinning of plants should be done before first irrigation to maintain a spacing of 5–10 cm between plants. Herbicides may also be applied for weed control as follows:

- Pre–plant herbicide - Fluchloralin 45 EC @ 0.75 kg/ha
- Pre–emergent herbicide - Oxyfluorfen 23.5 EC @ 0.15 kg/ha or Pendamithalin 30 EC @ 1.0 kg/ha

5.9. IRRIGATION

Irrigation requirement depends upon the parameters such as climate, soil moisture level and the variety used. Generally, 4–6 irrigations are given depending on the type of soil and climate. First irrigation should be given within 30–35 days from the date of sowing, second after 50–60 days, third after 70–80 days, fourth after 90–100 days, fifth after 105–110 days and sixth after 115–125 days.

5.10. HARVESTING

The Coriander is harvested in different growth stages as per its uses:

- **Green stage** - The Coriander may be harvested at any time before flowering, starting from the stage when it attains a height of about 20 cm.
- **Seed purposes** - The Coriander is harvested at full maturity stage for seed purposes. Coriander seeds should be harvested when 50 per cent seeds turn yellow. This crop matures in about 110–140 days. At maturity, seeds turn to yellowish green colour. After harvest, the crop should be dried under partial shade to retain the green colour and its aroma. When the plants are completely dried, the seeds should be separated by light beating with sticks. Seeds should be cleaned by winnowing or with the help of sieve or by vibrator and then should be graded. Fresh Coriander seeds should be dried in shade to retain seed colour and quality.

5.11. YIELD

The yield of Coriander varies according to the situation of cultivation:

- Rainfed condition - 500.0–800.0 kg/ha.
- Irrigated condition - 1,200.0–2,000.0 kg/ha.
- Green leaves - 1,000.0 kg/ha.

Clean gunny bags are used to pack Coriander seeds and they are stored in damp–free aerated store rooms.

5.12. VARIETIES

Gujarat Coriander-1, Gujarat Coriander-2, CS-287, DH-5, Rcr-20, RCr-41, Rcr-435, Rcr-436, Rcr-446, UD-20, Rajendra Swati (RD-44), CS-287, CO-1, CO-2, CO-3, Sadhana, Swati, Sindhu, Lam Sel. CS-2, Pusa Selection-360, UD-21, Cimpo 8-33, Morrocan, Gwalior No-5365, T-NP (D)-92, NP(D)-95, NP(J)-16, NP(J)-214M, NP(K)-45, Delhi local, Strain-85 etc.

5.13. PLANT PROTECTION

Insect pests

1. Coriander aphid

Scientific name - *Hyadaphis coriandri* (Das) and *Brevicoryne corianderi*

Aphid is most destructive insect of Coriander causing the loss of yield.

Fig. 17. Coriander aphid

Nature of damage

The nymphs and adults of these aphids suck plant sap from the leaves and tender shoots. The insect–affected leaves get twisted and the vitality of the plant is affected. The honeydew excreted by these insects gives out a glistening appearance of the canopy, which in turn gets covered with dark coating by the development of black mold. Severe infestation in early stages results in death of plants.

Control measures

- The resistant varieties of Coriander should be cultivated and late sowing should be avoided.
- The infested plant parts bearing aphid colonies should be clipped and destroyed by burning or burrowing under the soil.
- The cow dung ash should be dusted in the canopy early in the morning.
- The fungal pathogen, *Beauveria bassiana*, should be applied under humid condition.
- The aphid can be controlled by spraying any one of the insecticides given in Table 49.

Table 49. Insecticides used for controlling Coriander aphid

Sr. No.	Name of insecticide	Amount (per litre of water)
1.	Dimethoate 30 EC	2.0 ml
2.	Acephate 75 WP	0.75 g
3.	Cartap hydrochloride 50 SP	1.0 g
4.	Clothianidin 50 WDG	0.1 g

2. White fly

Scientific name: *Aleurodicus disperses, Aleuroth rixus floccosus, Aleyrodes proletella, Bemisia tabaci, Trialeurodes vaporariorum*

Fig. 18. White fly

Nature of damage

Whiteflies cause direct damage to plants by sucking plant sap. The viruses transmitted by white flies are among the most serious viruses on plants. Virus infection often results in total crop losses.

Control measures

- Dusting the crop with Malathion 5% or Carbaryl 50 WP @ 25.0 kg/ha in the white fly–affected field helps in controlling the insect.
- Use of bait trap prepared from different ingredients, as follows, is most effective in controlling the white fly.

Mollases	: 200 g
Malathion 50 EC	: 3.0 ml
Water	: 2.0 l

The poison bait traps should be taken in small earthen pots and kept in different places of the field.

- Spraying the crop with any one of the insecticides given in Table 50 is found effective in controlling white fly.

Table 50. Insecticides used for controlling the White fly of Coriander

Sr. No.	Name of insecticide	Amount (per litre of water)
1.	Neem oil 1500 ppm	3.0 ml
2.	Acephate 75 WP	0.75 g
3.	Acetamiprid 20 SF	1.0 g
4.	Malathion 50 EC	2.0 ml

Diseases

1. Wilt disease

C.O.: *Fusarium oxysporum* f. sp. *Corianderii*

Symptoms

The leaves of disease–affected plants become pinkish yellow to yellow. The disease can be easily recognized in the field by drooping of the terminal portions, followed by withering and drying up of leaves, eventually resulting in death. Severe infection in the early stage results in total failure of the crop.

Control measures

- Deep ploughing should be done during summer season.
- Crop rotation may also be followed.
- Sowing should be done after seed treatment with any one of the fungicides given in Table 51.

Table 51. Fungicides used for controlling the Wilt disease of Coriander

Sr. No.	Name of fungicide	Amount (per kg of seed)
1.	Carbendazim 50 WP	2.0–3.0 g
2.	Thiram 75 WP	2.0–3.0 g
3.	Tricoderma viridi	4.0–5.0 g

2. Powdery mildew

C.O.: *Erysiphe polygoni*

Symptoms

Small, white circular patches appear on young parts of stems and leaves, which increase in size, often coalesce to cover extensive areas of leaf surface. Affected leaves are reduced in size and distorted. In serve cases, the umbels dry up. White powdery growth appears on the leaves and buds during its primary stage. Premature sterility is also common. Seed formation may not take place in the disease–affected plants.

Fig 19. Powdery mildew disease of Coriander

Control measures

- Rouging of infected plants as soon as the infection is noticed in the field and subsequent burning is found as an effective method of controlling the disease.
- Foliar spray of onion extract (5 per cent) is effective in reducing severity.
- The disease can be controlled by spraying any one of the fungicides in the Table 52.

Table 52. Fungicides used for controlling the Powdery mildew of Coriander

Sr. No.	Name of fungicides	Amount (per litre of water)
1.	Sulphur 80 WP	2.0–3.0 g
2.	Azoxistrabin 50 WDG	1.0 ml
3.	Tridemorph 80 EC	1.0 ml

If needed two more sprays should be given at an interval of 15 d after first spray.

3. Stem gall

C.O.: *Protomyces macrospores*

The *stem gall* is a common disease of *Coriander,* which is an important spice crop of India and other countries.

Symptoms

The gall develops as a chemically induced swelling, arising from the surface of the leaf lamina, veins, mid-ribs, and petiole.

Symptom appears in the form of tumour–like swellings of leaf veins, leaf stalks, peduncles, stems as well as fruits. The swellings of the veins give a swollen hanging appearance to the leaves. Initially, the tumours are glossy, but later they rupture and become rough. The shape of Coriander seeds change as an effect of the disease.

Control measures

- Soil sterilization and use of healthy and disease–free seeds are the best remedies of the disease.
- Crop should be rotated for long term with other crops belonging to other family to reduce the primary inoculums.
- Sowing should be done after seed treatment with any one of the fungicides given in Table 53.

Table 53. Fungicides used for controlling the Stem gall of Coriander

Sr. No.	Name of fungicide	Amount (per kg of seed)
1.	Carbendazim 50 WP	2.0–3.0 g
2.	Thiram 75 WP	2.0–3.0 g
3.	Tricoderma viridi	4.0–5.0 g

- When the symptoms start appearing, spraying the crop with any one of the fungicides given in Table 54 is effective in controlling the disease. The spraying should be repeated at an interval of 20 d till the disease is completely controlled.

Table 54. Fungicides used for controlling the Stem gall of Coriander (for initial stage)

Sr. No.	Name of fungicide	Amount (per litre of water)
1.	Carbendazim 50 WP	1.0 g
2.	Mancozeb 75 WP	2.5 g

4. Bacterial leaf spot

C.O.: *Pseudomonas syringaepv.coriandicola*

Bacterial leaf spot is bacterial disease of Coriander.

Symptoms

Small (2–5 mm in diameter),angular–shaped, water–soaked leaf spot having greasy appearance appears on both sides of the infected leaf. As spots develop, they enlarge and turn dark brown in colour. The lesions increase in number with the progress of the disease, coalesce and can result in excessive leaf death.

Control measures

- Disease–free seeds collected from healthy crop should be used for sowing.
- The seeds should be treated in hot water (50°C for 20 min) before sowing.
- The disease–infected plants should be removed and destroyed by burning.
- The disease can be controlled by spraying the crop with any one of the bacteriacides in the Table 55.

Table 55. Bacteriacides used for controlling the Bacterial leaf spot of Coriander

Sr. No.	Name of bacteriacide	Amount (per litre of water)
1	Steptocycline (9:1) SP	0.1 g
2	Steptomycin 91.4SP + Tetracycline 4 SP	0.1 g

6 FENUGREEK

Scientific name - *Trigonella foenum-graecum*

Family - Fabaceae

Commercial part - Seed and fresh leaf

Indian names - *Bengali* : Methi; *Punjabi* : Methi; *Hindi* : Methi; *Urdu* : Methi; *Nepali* : Methi; *Tamil* : Menthiyam and venthayam; *Arabic* : Helba; *Malayalam* : Uluva and Ventayan.

Fenugreek (*Trigonella foenum-graecum*) is an annual plant in the family *Fabaceae*. According to history, the name Fenugreek or 'foenum–graecum' is derived from Latin for 'Greek hay'. The plant having small round leaves, is cultivated worldwide as a semi–arid crop, and is a common ingredient in dishes from the Indian Subcontinent. It is used as a condiment for flavouring of foods. It has got medicinal value, hence used as medicine. Another species of Fenugreek known as 'Kasuri' or 'Champa Methi' (*Trigonella corniculata*) belonging to same family is also cultivated in some States of India. The fresh tender green leaves of the Fenugreek plant, called methi or exuded, have a bitter–sweet flavour and are used as a regular green vegetable in Indian cuisine like dals and vegetables. Fenugreek is used as both an herb (the leaves) and as a spice (the seed).

Fenugreek seeds, an ancient spice, with a slightly bitter taste, are small, hard, yellowish–brown in colour, smooth and oblong–shaped.

6.1. ORIGIN AND DISTRIBUTION

Fenugreek is native to India and southern Europe. It is said that the Fenugreek seeds had been recovered from the tomb of Tutankhamen. The largest producer of Fenugreek in the world is India, where the major Fenugreek–producing States are Rajasthan, Gujarat, Uttarakhand, Uttar Pradesh, Madhya Pradesh, Maharashtra, Haryana, and Punjab. Rajasthan produces the lion's share of India's production, accounting for over 80 per cent of the nation's total Fenugreek output.

Fig 20. Fenugreek – 1. Plant. 2. Seed

6.2. CLIMATE

Fenugreek is a Rabi crop. It requires cool climate during vegetative growth and warm dry climate during maturity. It is tolerant to frost. It is a semi–arid crop cultivated worldwide and frequently used in many curries. This crop usually grows in cold season and is fairly tolerant to frost and very low temperature.

6.3. SOIL

Fenugreek can be grown on a wide variety of soils but, clayey loam is found relatively better for the cultivation of this crop. The optimum soil pH should be 6.0–7.0 for the growth and development of Fenugreek.

6.4. SOWING OF SEED

1. Time of sowing

The time of sowing is middle of October to first fortnight of November, but first fortnight of November is the best sowing time.

2. Seed rate

10.0–15.0 kg/ha

3. Method of sowing

The seeds are generally sown by broadcasting method, but line sowing can also be done at a distance of 25 cm. Seeds should be soaked in water for 6–8 hrs and dried in shade before sowing to hasten germination. Complete germination of seeds takes about 6–8 days time. A line–to–line spacing of 30 cm and plant–to–

plant spacing of 8–10 cm should be maintained while sowing or broadcasting of seeds.

6.5. MANURING

Manures

Compost or FYM: 10.0–15.0 t/ha

Fertilizers

Nitrogen (N)	: 40.0 kg/ha
Phosphorus (P_2O_5)	: 40.0 kg/ha

Method of application

Manures

The total amount of manures should be applied and mixed in soil at the time of land preparation.

Fertilizers

Basal dressing: Total amount of phosphorus and half amount of nitrogen

Top dressing: The remaining amount of nitrogen i.e., 20.0 kg at 30 DAS

6.6. INTERCULTURAL OPERATION

1. Weeding

The field of Fenugreek should be kept free from weeds particularly in early stages of crop growth. Intercultural operation during the early stage of plant growth minimizes weed competition. Two weedings need to be done to keep the field free from weed infestation as follows:

First weeding - 25–30 DAS

Second weeding - 40–50 DAS

The dense crop should be thinned, particularly in broadcasted crop, to maintain the plant distance about 8–10 cm. Herbicides may also be used for weed control. Pre–plant application of Fluchloralin @ 0.75 kg/ha supplemented with hand–weeding after the 50 d of sowing is proved to be highly effective for complete weed control.

6.7. IRRIGATION

Fenugreek needs sufficient moisture for its growth and development. Depending upon soil type and climate, 4–6 irrigations should be given to

Fenugreek crop. Pre–sowing irrigation should also be given, if moisture level of the soil is not optimum for seed germination. First irrigation should be given at the time of thinning and subsequent irrigation at an interval of 20–25 d. Water stress during pod development stage is detrimental and hence care should be taken to avoid water stress at pod and seed development stages.

6.8. HARVESTING

- **Green leaves** - Normally common methi leaves are nipped 2 cm above ground after 3–4 wk of sowing. In general 2–3 cutting are taken. When plant is uprooted, the crop is bundled and marketed.
- **Seeds** - Crop becomes ready for harvest in about 120–150 d. At the time of ripening or maturity, leaves and pods become yellowish and leaves start falling. The harvested plants are dried in threshing yard and threshed by biting with sticks or by trampling under the feet of bullocks. Seeds should be separated and cleaned by winnowing.

6.9. YIELD

1,800–2,000 kg/ha

6.10. VARIETIES

Kasuri, Pusa early bunching type, Prabha (NLM), Pusa early bunching No-14, EC-4911 Rajendra Kranti (RM-16), Guj. Methi-1, Methi No. 14, Methi No. 47, Rmt-1 (Rajasthan Methi), Co-1, UM-34, UM-35, Kasuri Selection etc.

6.11. USES

Fenugreek is used as a vegetable as well as a pulse. The leaves and young pods are used as vegetable and the seeds as condiments.

It also has some medicinal value. It prevents constipation, removes indigestion, stimulates the spleen and is appetizing and diuretic. The leaves are quite rich in protein, minerals and Vitamin C.

6.12. PLANT PROTECTION

Insect pests

1. Aphid

Scientific name - *Hyadaphis coriandri* (Das),

Nature of damage

Aphid remains on the plant in a colony and sucks the sap from the tender parts of the plant and affects the growth of plants adversely. Large populations cause curling, yellowing, and distortion of leaves and stunting of shoots. They can also produce large quantities of a sticky exuded known as honeydew, which often turns black with the growth of a sooty mold fungus. Aphids may transmit viruses from plant to plant.

Control measures

The aphid can be controlled by spraying any one of the insecticides given in Table 56.

Table 56. Insecticides used for controlling the Aphid of Fenugreek

Sr. No.	Name of insecticide	Amount (per litre of water)
1.	Dimethoate 30 EC	2.0 ml
2.	Acephate 75 WP	0.75 g
3	Imidachloropid 17.8 SL	0.2 ml

Diseases

1. Powdery mildew

C.O.: *Erisiphe polygoni*

Symptoms

White powdery patches appear on the lower and upper surface of the leaves and other parts of the plant. Powdery mildew disease appears usually in later stage of the crop and becomes serious when pod formation takes place.

Control measures

The disease can be controlled by spraying any one of the fungicides in the Table 57.

Table 57. Fungicide used for controlling the Powdery mildew of Fenugreek

Sr. No.	Name of fungicide	Amount (per litre of water)
1.	Sulphur 80 WP	2.0–3.0 g
2.	Azoxistrabin 50 WDG	1.0 ml
3.	Tridemorph 80 EC	1.0 ml

If needed, two more sprays should be given at an interval of 15 d after the first spray.

2. Downy mildew

C.O.: *Peronospora trigonellae*

Symptoms

Yellow patches on the upper surface of leaves appear in the infected plants and white cottony mycelium appears on the lower surface of leaves. This disease occurs during February and March.

Control measures

The disease can be controlled by spraying the crop with any one of the fungicides given in Table 58.

Table 58. Fungicides used for controlling the Downy mildew of Fenugreek

Sr. No.	Name of fungicide	Amount (per litre of water)
1.	Azoxistrabin 25 SP	1.0 ml
2.	Mancozeb 75 WP	2.5 g
3.	Flutraiazol 2.5 SP	2.0 g
4.	Propineb 70 WP	3.0 g

3. Bacterial blight

C.O.: *Pseudomonas syringae PV. Syringae*

Bacterial blight is a major disease of Fenugreek, which can severely reduce yield and seed quality. Bacterial blight is the first record of *Pseudomonas syringae* pv. *syringae* on Fenugreek in Australia.

Symptoms

Water–soaked, dark brown, angular spots are observed on the leaf surface and distributed along the veins. As symptoms progress, they cause leaf blight.

Control measures

- The seed treatment by hot water has considerable potential to reduce or even eliminate seed–borne inoculums.
- The infected plants and debris should be removed and destroyed.
- Crop rotation is recommended cultural control for bacterial blight.
- The disease can be controlled by spraying the crop with any one of the bactericides given in Table 59.

Table 59. Bactericides used for controlling the Bacterial blight of Fenugreek

Sr.No.	Name of bactericide	Amount (Per litre of water)
1.	Streptocycline (9:1) SP	0.1 g
2.	Bactrimycin SC	0.4 g
3.	Streptomycin 91.4 SP + Tetracyline 4 SP	0.1 g

7 CUMIN

Scientific name - *Cuminum cyminum*

Family - Apiaceae

Commercial part - Fruit

Indian names - *Bengali* : Safaid jira or Zeera; *Hindi* : Jira, Jeera, Zira or Safaid jeera or Zeera; *Gujarati* : Jiru or Jeeru; *Kannada* : Jeeriege; *Kashmiri* : Zyur; *Malayalam* : Jeerakam; *Marathi* : Jeregire; *Oriya* : Jira, Jeera; *Sindhi* : Zero; *Sanskrit* : Jiraka, Jira; *Tamil* : Ziragum or Jeeragam; *Telugu* : Jidakara, Jikaka.

Cumin (*Cuminum cyminum*) is a flowering plant in the family Apiaceae, native from the east Mediterranean to India.

The English 'Cumin' derives from the Old English cymen (or Old French Cumin), from Latin cuminum, which is the latinisation of the Greek kuminon) cognate with Hebrew kammon and Arabic kammun. Cumin (*Cuminum cyminum*) comes from Western Asia, where it has been cultivated since Biblical time.

Table 60. Nutritive value of Cumin

Energy	1,567 kJ (375 kcal)
Carbohydrates	44.24 g
Sugars	2.25 g
Dietary fibre	10.5 g
Fat	22.27 g
Protein	17.81 g
Water	8.06 g
Vitamin A equivalent	64 μg (8%)
Riboflavin (Vit. B_2)	0.327 mg (27%)
Niacin (Vit. B_3)	4.579 mg (31%)
Vitamin B_6	0.435 mg (33%)

Folate (Vit. B_9)	10 μg (3%)
Vitamin B_{12}	0 μg (0%)
Vitamin C	7.7 mg (9%)
Vitamin E	3.33 mg (22%)
Vitamin K	5.4 μg (5%)
Calcium	931 mg (93%)
Iron	66.36 mg (510%)
Magnesium	366 mg (103%)
Phosphorus	499 mg (71%)
Potassium	1788 mg (38%)
Sodium	168 mg (11%)
Zinc	4.8 mg (51%)

Percentages are relative to US recommendations for adults.
Source: USDA Nutrient Database

Fig. 21(a). Cumin seed

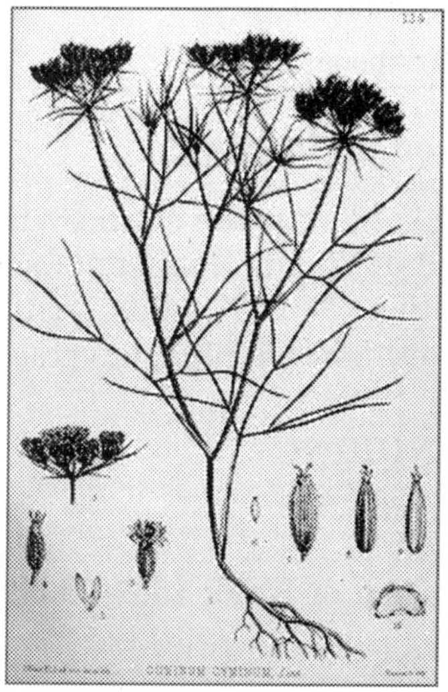

Fig. 21(b). Cumin Plant

7.1. CLIMATE

Cumin is a tropical plant. It grows well in sub–tropical climate too. It grows well in dry cool climate free from frost at 3,000 m MSL.

7.2. SOIL

Cumin can be cultivated in all types of soils. But well–drained sandy loam and medium soils are suitable for cultivation.

7.3. SOWING OF SEED

1. Time of Sowing

First week of December

2. Seed rate

12–15 kg/ha

3. Seed treatment

The seeds should be treated with any one of the seed treatment chemicals given in Table 61.

Table 61. Seed treatment chemicals - Cumin

Sr. No.	Name of seed treatment chemical	Amount (per kg of seed)
1.	Carbendazim 50 WP	2.0–3.0 g
2.	Captan 50 WP	2.0–5.0 g
3.	Thiram 75 WP	2.0–3.0 g
4.	Trichoderma viridi	4.0 g

4. Method of Sowing

The seeds may be sown by broadcasting or by drilling the seed in lines with seed drill at 30 cm. Soaking of seeds for 8 hours before sowing is helpful in good germination. Soaked seeds should be dried in shade to facilitate broadcasting. Sowing at higher depths, adversely affects the germination of seeds.

7.4. MANURING

Manures

Compost or FYM: 15.0–20.0 t/ha

Fertilizers

Nitrogen (N) : 30.0 kg/ha

Phosphorus (P_2O_5) : 15.0 kg/ha

Method of application

Manures

The total amount of manures should be applied at the time of land preparation and mixed thoroughly into the soil by ploughing followed by planking. Planking makes the soil levelled and compresses the soil, which helps in conservation of soil moisture.

Fertilizers

Basal dressing - Total amount of phosphorus and half amount of nitrogen.

Topdressing - The remaining amount of nitrogen should be applied as top dressing at 30 DAS.

7.5. INTERCULTURAL OPERATION

Weeds pose a problem in the production of Cumin. The Cumin field should be kept free from weeds for proper growth and development of plants.

Generally 2–3 hand weedings are required to keep the weeds under check. Light intercultural operation is beneficial in the drilled crop. First weeding and hoeing should be done at 30–40 DAS.

7.6. IRRIGATION

The crop grows at optimum soil moisture. The crop should be provided optimum water by providing irrigation depending on soil moisture content and weather condition. Crop requires 4–6 irrigations depending on the type of soil.

1st light irrigation should be given immediately after sowing and second irrigation should be given after 6–10 days from 1st irrigation. Subsequent irrigations should be given after 30, 45, 65 and 80 days from 1st irrigation. Irrigation at the time of flowering and fruit set are essential. At maturity stage, irrigation should be stopped.

7.7. HARVESTING

Generally, Cumin crop takes about 110–115 days to reach maturity. Crop becomes ready to harvest when plants turn yellowish–brown in colour. The Cumin is harvested when seed pods turn brown and when one seed head is dry enough to crack when pinched. Cumin is notorious for ripening unevenly. It is required to take care on ripening seed and plants should be harvested immediately. Harvesting should be done early in the morning by cutting/uprooting the whole plants. Once the first seed head is able to be pinched, the whole plant should be cut down and seeds should be allowed to dry in a paper bag. The Cumin seeds should be beaten out by threshing the dried plants with sticks. The seeds should further be dried to 10% moisture content, either by placing on mats or trays in the Sun or by using a drier if the conditions are too humid. The dried seeds should be winnowed using a traditional winnowing basket to remove the dirt, dust, leaves and twigs.

7.8. YIELD

600–700 kg/ha

7.9. VARIETIES

S-404, MC-43, Gujarat Cumin-1 (GC-1), GC-2, GC-3, RS-1, UC-198, RZ-19 etc.

7.10. USES

Cumin or jeera is a common ingredient in Indian kitchens. Cumin seeds have been extensively used in culinary preparations in the Indian subcontinent since ages. *Apart from adding flavour to a dish, it has got health benefits too.* It is a rich source of iron and hence very beneficial for anaemics as well as lactating mothers and pregnant women, who tend to need iron more than others.

7.11. PLANT PROTECTION

Insect pests

1. Aphid

Scientific name - *Myzus persicae* (Sulzer)

Aphid appears in early February and population remains static until the last week of February.

Nature of damage

Aphid remains on plant in a colony and sucks the sap from the tender parts of the plant, and adversely affects the growth of plants. Large populations cause curling, yellowing, and distortion of leaves and stunting of shoots. They can also produce large quantities of sticky exuded known as honeydew, which often turns black with the growth of a sooty mold fungus. Aphids may transmit viruses from plant to plant.

Control measures

The aphid can be controlled by spraying any one of the insecticides given in Table 62.

Table 62. Insecticides used for controlling the Aphid of Cumin

Sr. No.	Name of insecticide	Amount (per litre of water)
1.	Dimethoate 30 EC	2.0 ml
2.	Acephate 75 WP	0.75 g
3.	Clothianidin 50 WDG	0.1 g

2. Leaf eating caterpillar

Nature of damage

The leaf eating caterpillar causes damage to the foliage of plants reducing yield of the crop.

Control measures

The leaf eating caterpillar can be controlled by spraying any one of the insecticides given in Table 63 in the early stage of the crop.

Table 63. Insecticides used for controlling the Leaf eating caterpillar of Cumin

Sr. No.	Name of insecticide	Amount (per litre of water)
1.	Acephate 75 WP	0.75 g
2.	Thiomethoxam 25 WG	0.2 g
3.	Spinosad 45 SC	0.2 ml

Diseases

1. Fusarium wilt

C.O.: *Fusarium* sp

Symptoms

Infected plants show peculiar symptoms of dropping of tips and leaves, which leads to mortality of the entire plant. Attack of wilt is severe in younger plants.

Control measures

- Seeds collected from disease–free plots should only be used for sowing.
- Crop rotation and use of neem cake are helpful in checking spread of the fungus vis-à-vis disease.
- There is no chemical control for this disease.

2. Alternaria blight

C.O.: *Alternaria burnsii*

Symptoms

The blight–affected plants show very minute brownish necrotic spots, which later turn to blackish. Mostly diseased plants fail to produce seeds. If seeds are produced, they remain shrivelled, light in weight and dark in colour.

Control measures

- The crop should be kept free from weeds. For this, weeding should be done as per requirement.
- The seeds should be treated with seed treatment chemicals before sowing in the field.
- The disease can be controlled by spraying any one of the fungicides given in Table 64, four times at 10 d interval commencing from 40 days after sowing:

Table 64. Fungicides used for controlling the Alternaria blight of Cumin

Sr. No.	Name of fungicide	Amount (per litre of water)
1.	Copper oxychloride 50 WP	5.0 g
2.	Mancozeb 75 WP	2.5 g
3.	Carbendazim 50 WP	1.0 g

3. Powdery mildew

C.O.: *Erysiphe polygoni*

Affected plants in early stages show minute whitish spots on leaves, petiole, stem pedicel and seeds. In severe condition, it looks as the plants have been dusted with white powder. At later stages of attack seeds become white and shrivelled and light in weight.

Control measures

The disease can be controlled by spraying any one of the fungicides given in Table 65.

Table 65. Fungicides used for controlling the Powdery mildew of Cumin

Sr. No.	Name of fungicide	Amount (per litre of water)
1.	Sulphur 80 WP	2.0–3.0 g
2.	Azoxistrabin 50 WDG	1.0 ml
3.	Tridemorph 80 EC	1.0 ml

8 ANISE

Scientific name - *Pimpinella anisum*

Family - Apiaceae

Commercial part - Fruit and seed

Indian names - *Bengali* : Muhuri, Mitha jira; *Hindi* : Valaiti saunf or aawonf; *Gujarati* : Anisi, Sowa; *Kannada* : Sompu; *Malayalam* : Shombu; *Marathi* : Somp, Badishop; *Oriya* : Sop; *Punjabi* : Valaiti sounf; *Sanskrit* : Shatapusapa; *Tamil* : Shombu; *Telugu* : Kuppi soptu.

Anise (*Pimpinella anisu*), also called aniseed, is a flowering plant in the family *Apiaceae* or Carrot family. This herb is well known for its aromatic seeds. The taste of the seeds resembles that of fennel, which comes from the same family; or licorice, which comes from a different plant family. The Anise comes from English anis, from Latin anisum and from Greek anison.

8.1. BOTANICAL DESCRIPTION

- **Plant** - It is an herbaceous annual plant growing to 1 m (3 ft) or more tall and cultivated chiefly for its fruit, called aniseed.
- **Leaf** - The leaves are simple, 1–5 cm (⅜ - 2 inches) long and shallowly lobed, while leaves higher on the stems are feathery pinnate, and divided into numerous leaflets.
- **Flower** - The flowers are yellowish white in colour, small, approximately 3 mm (⅛ inches) in diameter and produced in dense umbels.
- **Fruit** - The fruit is oblong dry schizocarp, 3–6 mm (⅛–¼ inches) long, usually called 'aniseed' and greyish white in colour.

8.2. COMPOSITION

Moisture	: 9–13%
Protein	: 18%
Fatty oil	: 8–23%
Essential oil	: 2–7%
Starch	: 5%
N–free extract	: 22–28%
Crude fibre	: 12–25%

8.3. ORIGIN AND DISTRIBUTION

Anise is native to Egypt and the Mediterranean region and Southwest Asia. But it has since spread to Europe, northern Africa, Middle East and North America and now, it is grown all over the world. Ancient Egyptians and Romans used it as a spice. It has also been used throughout history for medicinal purposes and its fragrances.

Fig 22. (1) Anise plant (2) Seed

8.4. CLIMATE

Anise is a tropical plant. It grows well in sub–tropical climate too. It grows well in dry, cool climate, free from frost, at 3,000 m MSL.

8.5. SOIL

Anise likes poor, light, well–drained soil and full Sun (Kowalchik 1987). Anise should be planted in a location where it is blocked by wind because wind can damage this plant easily.

8.6. LAND PREPARATION

The land is brought to a fine tilth by tilling the land by tractor or power tiller 3–4 times followed by planking. The manures should be applied at the time of land preparation and mixed well into the soil by ploughing followed by planking.

8.7. SOWING OF SEED

1. Time of sowing

The sowing of seed is done in October–November in sub–tropical areas like North Indian plains and March-April in temperate areas like Europe.

2. Seed rate

5.0–10.0 kg/ha

3. Method of sowing

Anise is best propagated through seeds. The seeds should be sown by broadcasting at 1/2 inch deep in spring. The seeds may also be sown by seed drill in rows that are 2.5–3 ft apart. Anise seeds germinate more rapidly if sown near Coriander. It has been shown that the presence of Coriander improves the actual seed formation of the Anise plant. It is desirable to soak the seeds overnight in water before sowing and it improves the germination.

8.8. MANURING

Manures

Compost or FYM: 15.0–20.0 t/ha

Fertilizers

Nitrogen (N)	: 30.0 kg/ha
Phosphorus (P_2O_5)	: 15.0 kg/ha

Method of application

Manures

The total amount of manures should be applied at the time of land preparation and mixed thoroughly into the soil by ploughing followed by planking. Planking makes the soil levelled and compresses the soil, which helps in conservation of soil moisture.

Fertilizers

Basal dressing: Total amount of phosphorus and half amount of nitrogen.

Top dressing: The remaining amount of nitrogen should be applied as top dressing at 30 DAS.

8.9. INTERCULTURAL OPERATION

Weeds pose a problem in the production of Anise. The Anise field should be kept free from weeds during the initial period of growth for proper growth and development of plants. 2–3 weedings are recommended. Light intercultural operation is beneficial in drilled crop. First weeding and hoeing should be done at 30–40 DAS.

8.10. IRRIGATION

The crop grows at optimum soil moisture. A pre–planting irrigation is given just before sowing, and subsequently four irrigations from November to March are required to obtain a good high seed yield.

8.11. HARVESTING

Generally, Cumin crop takes about 110–115 days to reach maturity. Crop becomes ready to harvest, when plants turn yellowish brown. Harvesting should be done early in the morning by cutting/uprooting the whole plants. Harvested crop should be dried in the threshing yard.

The Cumin seeds are beaten out by threshing the dried plants with sticks. The seeds are then further dried to 10% moisture content, either by placing on mats or trays in the sun or by using a drier if the conditions are too humid. The dried seeds are winnowed using a traditional winnowing basket to remove the dirt, dust, leaves and twigs.

When the seeds are crisp and dry, they are rubbed between palms to separate the seed from the hull and are sieved to remove seeds from the husks. The cleaned seeds are stored in airtight containers.

8.12. YIELD

600–700 kg/ha

8.13. PLANT PROTECTION

Insect pests

Aniseeds have no serious pest problems. Aniseed oil is said to repel insects.

Diseases

Aniseeds have no serious disease problems.

8.14. USES

The aniseeds are used whole or ground. It is mainly used as a flavourant in culinary, household, cosmetic and medicines. The fruit of Anise, commercially called aniseed, is widely used for flavouring curries, bread, soups, baked goods, dried figs, deserts, cream cheese, pickles, egg dishes, and non–alcoholic and alcoholic beverages. The essential oil is valuable in perfumery. The oil is used for production of anethole and sometimes as sensitizer for bleaching colours in photography. The other functional properties are anti–bacterial, anti–fungal, anti–oxidant, stimulant, carminative and expectorant.

The aniseeds have some health benefits as follows:

- In India, aniseeds are often chewed after a meal, both as a digestive and to sweeten the breath.
- Aniseeds are attributed to have anti–parasitic properties. Thus, it is used in home remedies to treat fungal infections and to get rid of lice etc.
- Aniseeds have other functional properties such as anti–bacterial and anti–oxidant nature.
- Aniseeds have numerous purported medicinal properties, mainly as a digestive aid.
- Anise oil, extracted from aniseeds, has also been used in treating flatulence and colic in children.

9 FENNEL

Scientific name - *Foeniculum vulgare*

Family - Apiaceae (formerly the Umbelliferae)

Commercial part - Bulbs, stalks, leaves and seeds

Fennel plant is a hardy, perennial, umbelliferous herb, with yellow flowers and feathery leaves. It is used as condiment and culinary spice.

9.1. ORIGIN AND DISTRIBUTION

Fennel is indigenous to the shores of the Mediterranean but has become widely naturalized in many parts of the world, especially on dry soils near the seacoast and on riverbanks. Fennel generally known as Saunf (*Foeniculum vulgare*) belongs to Apiacae family, which is mainly cultivated in Gujarat, Rajasthan and Uttar Pradesh.

9.2. BOTANICAL DESCRIPTION

- **Plant** - It is hardy, perennial umbelliferous herb. It is erect, glaucous green, and grows to heights up to 2.5 m, with hollow stems.
- **Leaf** - The leaves are feathery and grow up to 40 cm long. They are finely dissected, with the ultimate segments filiform (threadlike), about 0.5 mm wide.
- **Flower** - The flowers are produced in terminal compound umbels 5–15 cm wide, each umbel section having 20–50 tiny yellow flowers on short pedicels.
- **Fruit** - The fruit is a dry seed from 4–10 mm long.

Fig 23. Fennel

9.3. CLIMATE

Fennel requires cool dry climate. Frost prone areas should be avoided as this crop is susceptible to frost. Rainfall during maturity spoils the colour and reduces the quality of Fennel seeds. Generally, Fennel is cultivated as Kharif as well as Rabi crop.

9.4. SOIL

Well–drained loamy soil and black cotton soils are suitable for cultivation of this crop.

9.5. LAND PREPARATION

The land should be brought to a fine tilth by ploughing 2–3 times with tractor or power tiller followed by planking. Manures and fertilizers should be mixed properly in the soil at the time of land preparation.

9.6. SOWING OF SEED

1. Time of sowing

Kharif crop - Middle of June is suitable time for sowing of seeds in nursery.

Rabi crop - Last week of August to 1st week of September is suitable time of sowing.

2. Seed rate

Kharif crop - 2–2.5 kg/ha

Rabi crop - 2.5–3.0 kg /ha

3. Seed treatment

The seeds should be treated with any one of the seed treatment chemicals given in Table 66.

Table 66. Seed–treatment chemicals for the seed treatment of Fennel

Sr. No.	Name of seed–treatment chemical	Amount (per kg seed)
1.	Carbendazim 50 WP	2.0–3.0 g
2.	Captan 50 WP	2.0–5.0 g
3.	Thiram 75 WP	2.0–3.0 g
4.	Trichoderma viridi	4.0 g

4. Method of sowing seed

The seed should be sown by broadcasting or line sowing by placing by hand. Beds should be covered with plant wastes or straws or banana leaves for up to 12 d with a view to get good germination. Tender seedlings should be protected from direct sunlight, by erecting temporary shed over the bed. When seedlings are free from the danger of sunlight, temporary shed should be lifted. The manuring schedule in nursery bed should be as follows:

Kharif season - The manuring schedule in nursery bed (Area: 100 sq m):

FYM : 250.0 kg

Nitrogen (N) : 0.60 kg

Phosphorus (P_2O_5) : 0.30 kg

Rabi season - The manuring schedule in nursery (Area: 150–200 sq m):

FYM : 375.0–500.0 kg

Nitrogen (N) : 1.0–1.2 kg

Phosphorus (P_2O_5) : 0.5–0.06 kg

The whole quantity of FYM should be mixed in soil at the time of land preparation and whole dose of nitrogen and phosphorous should be applied as basal dose. Supplementary dose of urea can be applied three weeks after the germination of seeds, if seedlings are very weak.

9.7. TRANSPLANTING

The Fennel seedlings are suitable for transplanting in the main field when they are at about 15 cm height.

- **Kharif season**: Fennel seedlings should be transplanted at a spacing of 90 x 60 cm or 100 x 60 cm. Suitable time for taking up transplanting of Fennel is around 15th August.
- **Rabi season**: Fennel seedlings should be transplanted adopting a spacing of 45 x 10 cm.

9.8. MANURING

Kharif season

Manures

Compost or FYM: 20.0 t/ha

Fertilizers

Nitrogen (N)	: 100.0 kg/ha
Phosphorus (P_2O_5)	: 60.0 kg/ha

Method of application

Manures

The total quantities of manures should be applied at the time of land preparation and mixed in the soil by ploughing followed by planking.

Fertilizers

Basal dressing: Total amount of phosphorus and 40.0 kg nitrogen

Top dressing: Remaining 60 kg nitrogen should be top dressed in two equal splits of 30 kg each at an interval of 30 days and 60 days (at the time of earthing up) after transplanting.

Rabi season

Manures

Compost or FYM: 20.0 t /ha

Fertilizers

Nitrogen (N)	: 90.0 kg/ha
Phosphorus (P_2O_5)	: 45.0 kg/ha

Method of application

Manures

The total quantities of manures should be applied at the time of land preparation and mixed in the soil by ploughing followed by planking.

Fertilizers

Basal dressing - Total amount of phosphorus and 45.0 kg of nitrogen

Top dressing - Remaining 45 kg nitrogen should be applied as top dressing in two equal splits of 22.5 kg each at an interval of 30 d and 60 d (at the time of earthing up) after transplanting.

9.9. INTERCULTURAL OPERATION

1. Weeding

Weeds pose a problem in the growth and production of Fennel. So, land should be kept free from weeds, particularly in the early stages of the growth of the crop. First intercultural operation/hoeing should be done 20–25 d after transplanting. Intercultural operations should be repeated twice based on the need, at an interval of 20–25 d.

2. Earthling up

Earthling up should be done after 60–75 d from transplanting to check the lodging of plants.

9.10. IRRIGATION

Kharif season

Fennel is a long duration crop. The crop should be irrigated during dry spell in monsoon and after the monsoon as per the need of the crop. Generally, 10–12 irrigations should be given to Fennel crop depending on soil and climate. First irrigation should be given on the day of transplanting and subsequent irrigations should be given at an interval of 15–20 d up to February. Irrigation at the time of seed development is critical. So measures should be taken to provide irrigation at the seed development stage.

Rabi season

Generally, 8–10 irrigations are given to Rabi Fennel crop depending on soil and climate. First irrigation should be given on the day of transplanting and subsequent irrigations should be given at an interval of 15–20 d up to February. Irrigation at the time of seed development is critical. So measures should be taken for providing irrigation at the seed development stage.

9.11. HARVESTING

- **Leaf** - The leaves should be harvested when plants become established and start growing well. Each time a few leaves should be taken to minimize the harm to the plant.
- **Seed** - Harvesting of Fennel seeds should be done in the fall when the flowers turn brown, and they should be left to dry on the plant.
- **Bulb** - The bulb should be harvested once it gets the size of a small tennis ball. If you leave it to grow larger, the plant will likely bolt to seed and the taste of the bulb will almost immediately become bitter.

Harvesting should be done by cutting the umbels as and when they reach physiological maturity (when seeds are fully filled up and become green in colour). After cutting the umbels, they should be dried in partial shade. While drying, care should be taken for better aeration and umbels should be over turned, if necessary. The seeds should be separated by beating the dry umbels with sticks followed by winnowing. Grading is done with the help of sieve or vibrator. For seed purpose, fully matured umbels are harvested.

9.12. YIELD

- Kharif season - 1,800–2,000 kg/ha
- Rabi season - 1,500–1,800 kg/ha

9.13. VARIETIES

PF-35, Guj. F-1, Guj.F-2 etc.

9.14. PLANT PROTECTION

Insect pests

1. Aphids

Scientific name - *Hyadaphis foeniculi* (Passerini)

The aphid is most destructive insect.

Nature of damage

Aphid remains on plant in a colony and sucks the sap from the tender parts of the plant and affects the growth of plants adversely. Large populations cause curling, yellowing, and distortion of leaves and stunting of shoots. They can also produce large quantities of a sticky exuded known as honeydew, which often turns black with the growth of a sooty mold fungus. Aphids may transmit viruses from plant to plant.

Control measures

The aphids can be controlled by spraying any one of the insecticides given in Table 67.

Table 67. Insecticides used for controlling the Aphid of Fennel

Sr. No.	Name of insecticide	Amount (per litre of water)
1.	Dimethoate 30 EC	2.0 ml
2.	Acephate 75 WP	0.75 g
3.	Cartap hydrochloride 50 SP	1.0 g
4.	Clothianidin 50 WDG	0.1 g

Diseases

1. Damping off seedlings

C.O.: *Phytophthora syringae*

Symptoms

Infected seedlings are toppled on the ground surface. (Post–emergence damping off).

Control measures

Drenching of nursery beds with copper oxychloride @ 0.25% should be done to control this disease effectively.

2. Collar rot

C.O.: *Sclerotium sclerotiorum*

Symptoms

This disease appears in those plots where water stagnation near the plant stem is more. Collar portion of the plant starts decaying and the plant turns yellow in colour and dies off later.

Control measures

- The water stagnation in the field should be avoided.
- Drenching by 1% Bordeaux mixture or spraying the crop with any one of the fungicides given in Table 68 should be done to control the disease.

Table 68. Fungicides used for controlling the Collar rot of Fennel

Sr. No.	Name of fungicide	Amount (per litre of water)
1.	Copper oxychloride 50 WP	5.0 g
2.	Captan 50 WP	2.5 g
3.	Copper hydroxide 77 WP	2.0 g
4.	Carbendazim 50 WP	1.0 g

3. Ramularia blight

C.O.: *Ramularia foeniculi*

Symptoms

At the beginning of this disease, small ashy spots are observed on the leaves. These spots get enlarged and turn into brown colour. In case of severe attack, whole plant turns to brown colour, resulting in drying up of the plant.

Control measures

Spraying the crop with any one of the fungicide given in Table 69 is found to be effective in controlling this disease:

Table 69. Fungicides used for controlling the Ramularia blight of Fennel

Sr. No.	Name of fungicide	Amount (per litre of water)
1.	Copper oxychloride 50WP	5.0 g
2.	Mancozeb 75 WP	2.5 g
3.	Copper hydroxide 77WP	2.0 g
4.	Carbendazim 50WP	1.0 g

10 DILL

Scientific name - *Anethum graveolens* L

Syn - *Peucedanum graveolens* (L.) C. B. Clark

Family - Apiaceae

Commercial parts - Seeds, herb, oil

Indian names - *Bengali* : Sowa; *Hindi* : Sowa; *Gujarati* : Surva; *Kannada* : Sabasige; *Kashmiri* : Sor; *Malayalam* : Sathakuppa; *Marathi* : Surva, Shepu; *Punjabi* : Sowa; *Sanskrit* : Satapushpi; *Tamil* : Sathakuppi, Sompa; *Telugu* : Sabasiege; *Urdu* : Sowa.

Dill is an annual herb that is a member of the parsley family. Fresh and dried Dill leaves (sometimes called 'Dill weed' to distinguish it from Dill seed) are used as herbs, mainly in Poland, Finland, Sweden, the Baltic, Russia, and in Central Asia. Dill seeds are used as spice. This herb is grown for its bitter seeds that are used to make Dill pickles. Dill is also known as Dill weed. This herb is native to southwest Asia. It has many uses in the kitchen including as an ingredient in soups and stews. Dill, a popular annual plant, has bluish–green stems that contrast with finely divided, yellow-green coloured, plume-like leaves and yellowish flowers. Dill grows about 2–3 ft high.

10.1. ORIGIN AND DISTRIBUTION

European Dill (*Anetheum graveolens*) is indigenous to Europe and is cultivated in England, Germany, Romania, Turkey, USA and Russia. The Indian Dill (*Anetheum sowa*), a native of Northern India is bolder than the European Dill. It is cultivated as a cold weather crop in many parts of India. The seeds are stronger and more flavourful than the leaves and are most commonly associated with the cuisines of Scandinavia and Germany.

Fig 24. Dill

Table 70. Nutritive value of Dill

Dill, fresh	
Nutritional value per 100 g (3.5 oz)	
Energy	180 kJ (43 kcal)
Carbohydrates	7 g
Dietary fibre	2.1 g
Fat	1.1 g
Protein	3.5 g
Vitamin A	7717 (154%) IU
Thiamine (Vit. B_1)	0.1 mg (9%)
Riboflavin (Vit. B_2)	0.3 mg (25%)
Niacin (Vit. B_3)	1.6 mg (11%)
Pantothenic acid (B_5)	0.4 mg (8%)
Vitamin B_6	0.2 mg (15%)
Folate (Vit. B_9)	150 μg (38%)
Vitamin B_{12}	0 μg (0%)
Vitamin C	85 mg (102%)
Calcium	208 mg (21%)

Iron	6.6 mg (51%)
Magnesium	55 mg (15%)
Manganese	1.3 mg (62%)
Phosphorus	66 mg (9%)
Potassium	738 mg (16%)
Sodium	61 mg (4%)
Zinc	0.9 mg (9%)
Copper	0.14 mg (7%)

Percentages are relative to US recommendations for adults.
Source: USDA Nutrient Database

10.2. CLIMATE

Dill is grown as Rabi crop or winter season crop for commercial seed production in India. Bright sunshine is favourable for cultivation of Dill. Dill should be planted in a location that receives at least 6–8 h of direct sunlight daily. It can resist frost to a limited scale during vegetative stage. Dry and relatively high temperatures are desirable during seed production.

10.3. SOIL

Dill grows best in a well–drained, slightly acidic soil that is rich in organic matter.

10.4. PROPAGATION

Dill is most often grown from seeds. Dill is easily grown from seed sown in the garden in spring after all frost has passed away. Dill is best grown in rows 24 inches wide and 24 inches apart. Seeds should be sown throughout the season to ensure a continual harvest.

10.5. SOWING OF SEED

1. Time of sowing

October–November; but, the best time of sowing is mid–October to mid–November.

2. Seed rate

3.0–4.0 kg/ha

3. Seed treatment

The seeds should be treated with any one of seed treatment chemicals given in Table 71.

Table 71. Seed–treatment chemicals used for the seed treatment of Dill

Sr. No.	Name of seed–treatment chemical	Amount (per kg seed)
1.	Carbendazim 50 WP	2.0–3.0 g
2.	Captan 50 WP	2.0–5.0 g
3.	Thiram 75 WP	2.0–3.0 g
4.	Trichoderma viridi	4.0 g

4. Method of sowing

The seeds are generally sown in well–prepared soil and then covered by planking. It can also be sown by seed drill at spacing of 25–30 cm between row and a depth of 2–3 cm.

When the seedlings are about six inches tall, thinning is done to allow 12–15 inches between each plant.

10.6. INTERCULTURAL OPERATION

1. Weeding

Wedding should be done in time to keep the field free from weeds. The first weeding should be done at 21–28 DAS and subsequent weeding should be done as per the need. Herbicide (e.g., Thiobencarb 50 EC @ 800 ml as pre–emergence) is also found to be effective in controlling the weeds.

2. Thinning

Thinning should be done at 21 DAS to keep the distance in between the plants about 8–10 cm along the rows.

3. Mulching

Mulching is a process, which helps in conserving soil moisture and keeping the weeds under check. Mulching may be done by spreading straw, tree leaves, plastic sheets etc. The weeds that are uprooted may also be used as mulch. Protection should be provided to Dill plant from the more intense rays of bright sunlight.

10.7. IRRIGATION

Irrigation should be given as per the need of the plant depending on the weather condition. Irrigation should be given in such a manner that keeps the soil moist, but not wet. First irrigation should be given immediately after sowing the seeds to ensure good germination of seeds. Subsequent irrigation should be given at flowering and seed development stage. Irrigation should be given in the cooler hours of the morning or in the evening.

10.8. HARVESTING

Dill crop takes about 70–80 d for flowering.

Dill plants provide two distinct crops– leaves and seeds.

1. Dill leaf

Green dill foliage can be harvested anytime during the growing season until the umbrella–like flower clusters are open. Dill foliage can be dried by hanging the plant upside down in a warm, breezy place out of direct sunlight. Leaves, usually referred to as Dill weed, can be harvested once the plant is fairly established. Dill crop takes about 70–80 d for flowering.

2. Dill seed

The crop becomes ready for harvest by middle of March. To harvest the seeds, the flower stalks should be cut just before seeds begin to ripen and turn a tan colour. The stalk should be hanged upside down in a warm, well–ventilated room away from direct light. A small paper bag should be placed around the flower heads, fastened to the stalks. A few holes should be poked in the sides of the bag for air circulation. As the seeds ripen, they will drop and can be collected at the bottom of the bag. Seeds can be stored up to a year in air–tight containers as long as they are kept away from heat and bright light. Seeds must be very dry before they are stored; if any signs of moisture appear in the container shortly after storage, remove the seeds and dry them further. As Dill loses its flavour quickly, it is best to use it fresh, as soon after picking as possible.

10.9. YIELD

6.0–9.0 q/ha

10.10 USES

- **Green herb** - The green herb is used as a pot–herb and as a flavouring agent. Dill stems and umbels are used in Dill pickles and for flavouring soups.
- **Seed** - Dill seeds are used, both whole and ground, as a condiment in soups, salads, processed meals, sausages, spicy table sauces, salads, saucerkrant and particularly in Dill pickles. The essential oil is also used in the manufacture of soaps etc.

10.11. PLANT PROTECTION

Dill plants are not affected by serious insect pest except leaf–eating caterpillar and powdery mildew.

11 CLOVES

Scientific name - *Syzygium aromaticum* (L.) Merrill & Perry

Syn - *Caryophyllus aromaticus* L.

Eugenia aromatica (L.) Baill.

Eugenia caryophyllata Thunb.

Eugenia caryophyllus (Spreng.) Bullock and S. G. Harrison

Family - Myrtaceae

Commercial parts - The dried flower buds

Indian names - *Bengali* : Labanga; *Hindi* : Laung; *Gujarati* : Lavang; *Kannada* : Lavanga; *Malayalam* : Grambu; *Marathi* : Luvang; *Oriya* : Labanga; *Punjabi* : Laung; *Sanskrit* : Lavanga; *Tamil* : Kirambu, Lavangam; *Telugu* : Lavangalu.

Cloves are the aromatic dried flower buds of a tree, *Syzygium aromaticum* in the family Myrtaceae. The clove is a flower bud from an aromatic evergreen tree that grows in the tropics of Asia and South America. The name clove is derived from the Latin, clavus, meaning nail, due to the possible nail–shape of the dried bud. The Chinese first documented the use of fragrant clove in 207 B.C.

Cloves are harvested when 1.5–2 cm long, and consist of a long calyx, terminating in four spreading sepals, and four unopened petals which form a small ball in the center. The clove is the aromatic, fully grown but unopened dried flower buds of *Syzygium aromaticum* (also known as *Eugenia caryophyllus*). It is one of the ancient valuable spices of the orient, holding a unique position in the international trade.

11.1. COMPOSITION

Table 72. Composition of Cloves

Moisture	5.4%
Protein	6.3%
Volatile oil	13.2%
Non-volatile ether extract (fat)	15.5%
Crude fibre	11.1%
Carbohydrates	57.7%
Mineral matter	5.0%
Calcium	0.7%
Phosphorus	0.11%
Iron	01%
Sodium	0.25%
Potassium	0.2%
Vitamin B1	0.11 mg/100 g
Vitamin B2	0.04%
Niacin	1.5%
Vitamin C	80.9 mg/100 g
Vitamin A	175 I.U
Calorific value (food energy)	430 calories/100 g

11.2. BOTANICAL DESCRIPTION

- **Plant -** It is an evergreen herb that grows to a height ranging from 8–12 m.
- **Leaf -** The leaves are large in size.
- **Flower -** Sanguine flowers are in numerous groups of terminal clusters. Initially, the flower buds are of pale colour and gradually become green, after which they develop into bright red colour, when they are ready for collecting.

11.3. ORIGIN AND DISTRIBUTION

Cloves are native to the Molucca islands in Indonesia and India and are used as a spice in cuisines all over the world. Cloves are harvested primarily in Indonesia, India, Madagascar, Zanzibar, Pakistan and Sri Lanka. It originated in Moluccas, the so called 'Spice Island' in Indonesia. The islands of Zanzibar and Pemba (now parts of Tanzania), Indonesia, Madagascar, Sri Lanka and Malaysia are the important producers of clove. Clove was first introduced to India around 1800 A.D. by the East India Company in its 'spice garden' in Courtallam, Tamil Nadu. Now, the important clove–growing regions in India are Nilgiris, Tirunelveli, Kanyakumari, Nagercoil and Ramanathapuram districts of Tamil Nadu; Kozhikode, Kottayam, Kollam and Thiruvananthapuram districts of Kerala, and South Kanara district of Karnataka.

Cloves are harvested when the flowers turn from green to purple. The ripened flower buds are collected and sun–dried in tropical regions.

Clove grows best in rich loamy soils in the wet tropics. Clove can also grow in heavier red soils, but in either case, needs good drainage. Cloves are propagated by seeds or by cuttings. The seeds can be directly planted, or soaked in water overnight to remove the outer lint.

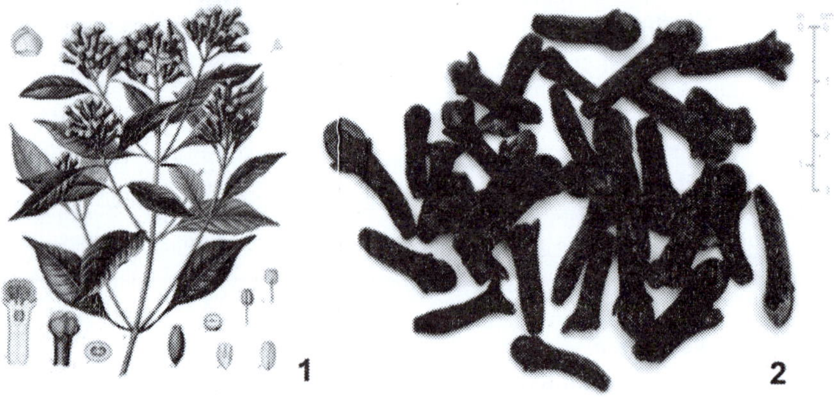

Fig 25. 1. Clove twing 2. Clove

11.4. CLIMATE

Clove is a tropical plant and requires warm humid climate. Clove thrives in all situations ranging from sea level up to an altitude of 1,000 m. Under Indian conditions, it is best suited for mixed cropping in either coconut or arecanut gardens or in coffee estates.

11.5. SOIL

Deep loam soil with high humus content, found in the forest region, is best suited for its cultivation. It grows satisfactorily on laterite soil, loam and rich black soil having good drainage.

11.6. PROPAGATION

Clove is propagated through seed obtained from ripened fruit, popularly known as 'mother of clove'. Fruits are taken from trees aged more than 15 yr and with regular yielding nature.

Seeds should be placed flat at a depth of about 2.5 cm with a spacing of 12–15 cm. Germination commences in about 10–15 d and completes by about 45 d. The slender and delicate seedlings grow very slowly.

11.7. PLANTING

1. Time of planting

June–July

2. Method of Planting

The slender and delicate plants should be planted in a pit measuring 75 x 75 x 75 cm, dug at a spacing of 6–7 m. The pits should be partially filled with compost, green leaf or cattle manure and covered with top soil. Clove prefers partial shade.

11.8. MANURING

The judicious manuring is found good for getting higher yield. The amount of manures varies with the age of the plant as follows:

Table 73. Manures schedule for Clove

Sr. No.	Age of plant	Manures (kg/plant)	Amount of nutrient (g/plant)		
			Nitrogen (N)	Phosphorus (P_2O_5)	Potassium (K_2O)
1.	1 yr	15.0	20	20	60
2.	2 yr	30.0	40	35	100
3.	15 yr	50.0	300	250	750

In addition to this, 50 g of each *Azospirillum* and *Phosphobacterium* should be applied one month after manuring.

Method of application

Manures

Manures should be applied in the month of May–June, with the commencement of monsoon.

Fertilizers

Fertilizers should be applied in two equal split doses i.e., first during May–June and other during September–October.

Manures and fertilizers need to be applied in the trench that is dug around the tree about 50–160 cm away from the base depending upon the age.

11.9. INTERCULTURAL OPERATION

The land should be kept free from weeds. For this, weeding should be done at regular interval. Dead and diseased shoots should be removed once or twice a year.

11.10. IRRIGATION

Irrigation should be given in the initial stage of the crop. Afterwards, irrigation should be given as per the demand of the crop for the proper growth and yield.

11.11 HARVESTING

Clove tree begins to yield from the seventh year of planting and full bearing stage is attained after 15–20 yr. The flowering season is September–October in the plains and December–February at high altitudes.

The clove buds should be harvested when the buds are fully developed and the base of the calyx has turned from green to pink colour.

The buds, after separation from the stalks, should be spread evenly on mats or cement floors to dry in the Sun. Fully–dried buds develop the characteristic dark brown colour and are crisp.

11.12 YIELD

2.0 kg per plant or 400.0 kg/ha

11.13. USES

The bud and stem of the dried clove is widely used as a spice and food flavouring. The oil extracted from the plant, leaves, flower buds, and fruit itself is used in herbal remedies and some dental practices.

11.14. PLANT PROTECTION

Insect pests

1. Stem borer

Scientific name - *Sahyadrassus malabaricus*

Stem borer is the most important insect pest of clove.

Nature of damage

The caterpillars bore into the main stem resulting in immediate drying up of the plant above the point of attack and causing the death of the plant ultimately.

Control measures

- Clean cultivation (i.e., timely weeding, destruction of insect affected parts of plant etc) is most effective method of reducing the incidence of infection.
- Stem borer can be controlled by spraying the solution of any one of the insecticides given in Table 74.

Table 74. Insecticides used for controlling the Stem borer of Clove

Sr. No.	Name of insecticide	Amount (per litre of water)
1.	Thiomethoxam 25 WG	0.2 g
2.	Quinalphos 25 EC	2.0 ml
3.	Acephate 75 WP	0.75 g
4.	Imidachloropid 17.8 SL	0.2 ml

2. Scales

Scientific name - *Lecanium psidii*

The scales are one of the serious sucking type insects of clove. It is serious insect in the nursery.

Nature of damage

The scales suck the sap from the leaves and tender shoots. Young seedlings, if attacked, are killed soon.

Control measures

- Lady bird beetle such as *Kainochoras nigritus* is found effective in controlling scale biologically.
- The insect can be controlled by spraying the crop with any one of the insecticides given in Table 75.

Table 75. Insecticides used for controlling the Scales of Clove

Sr. No.	Name of insecticide	Amount (per litre of water)
1.	Acephate 75 WP	0.75 g
2.	Imidachloropid 17.8 SL	0.2 ml
3.	Thiomethoxam 2	0.2 g

3. Mealy Bugs

Scientific name - *Planococcus* sp, *Psuedococus* sp

Mealy bug is also a sucking type insect.

Nature of damage

Mealy bug sucks the sap from the tender shoots. The insect–affected parts of the plant dry up gradually.

Control measures

- The mealy bug can be controlled by spraying the plant simply with cold water.
- Lady bird beetle such as *Cryptolisus montochhari* is found very much beneficial for controlling the mealy bug.
- The insect can be controlled by spraying the crop with any one of the insecticides given in Table 76.

Table 76. Insecticides used for controlling the Mealy bugs of Clove

Sr. No.	Name of insecticide	Amount (per litre of water)
1.	Acephate 75WP	0.75 g
2.	Imidachloropid 17.8 SL	0.2 ml
3.	Thiometon 25 EC	2.0 ml
4.	Malathion 50 EC	2.0 ml

Diseases

1. Leaf rot

C.O.: *Cylindrocladium quinquiseptatum*

The leaf spot disease appears both in nursery and in the main field at young and mature stage respectively.

Symptoms

Infection starts as dark spots at the leaf margin and spreads, sometimes, with no definite pattern. Rotting may be in the whole leaf or at the tip resulting in defoliation.

Control measures

The disease can be controlled by spraying the plant with any one of the fungicides given in Table 77.

2. Leaf spot, Twig blight and Flower bud shedding

C.O.: *Colletotrichum gleosporioides*

Table 77. Fungicides used for controlling the Leaf rot of Clove

Sr. No.	Name of fungicide	Amount (per litre of water)
1.	Copper oxychloride 50 WP	5.0 g
2.	Carbendazim 50 WP	1.0 g
3.	Metalaxyl 35 WS	0.5 g

Symptoms

Necrotic spots of variable size and shapes are noticed on the leaves. Severely–affected leaves wither, droop down and dry up. In the nursery, seedlings die–back symptoms are seen. Twigs are infected as the symptoms extend from the leaves through petiole.

Control measures

The disease can be controlled by spraying the plant with any one of the fungicides given in Table 78, at an interval of 15 d depending on the severity of the disease.

Table 78. Fungicides used for controlling the Leaf spot, twig blight and Flower bud shedding of Clove

Sr. No.	Name of fungicide	Amount (per litre of water)
1.	Mancozeb 75 WP	2.5 g
2.	Carbendazim 50 WP	1.0 g
3.	Copper oxychloride 50 WP	5.0 g
4.	Thiophenate methyl 70 WP	1.0 g
5.	Metalaxyl 35 WS	0.5 g

4. Die back

C.O. *Crytosporlla eugeniae*

Symptoms

The young seedlings and grown–up trees are affected by die back disease. The disease–affected leaves rot and fall leading, occasionally, to total defoliation. The twigs also rot starting from tips and proceed downwards resulting in drying up of branches.

Control measures

The disease can be controlled by spraying the plant with any one of the fungicides given in Table 79.

Table 79. Fungicides used for controlling the Die back of Clove

Sr. No.	Name of fungicide	Amount (per litre of water)
1.	Mancozeb 75 WP	2.5 g
2.	Carbendazim 50 WP	1.0 g
3.	Copper oxychloride 50WP	5.0 g
4.	Thiophenate methyl 70 WP	1.0 g
5.	Metalaxyl 35WS	0.5 g

12 CINNAMON

Scientific name - *Cinnamomum verum*

Family - Lauraceae

Commercial part - Bark, leaf

Indian names - *Hindi* : Dalchini, Darchini; *Bengali* : Dalchini; *Gujarati* : Dalchini; *Kannada* : Lavangapattai; *Malayalam* : Karuvapatta; *Marathi* : Dalchini; *Oriya* : Dalchini; *Punjabi* : Dalchini; *Sanskrit* : Darushila; *Tamil* : Karuvapattai, Sannalavangapattai; *Urdu* : Dalchini.

Cinnamon or 'sweet wood' is the earliest–known spice in India. It is native of Sri Lanka and Malabar Coast of India. It is grown in the Naga Hills of Assam, Coastal hills of Karnataka and the Western Ghats. Cinnamon is an evergreen tree, dried inner bark of which is used as a spice. Cinnamon is naturally seen in the forests of Western Ghats and commercially cultivated in Kerala, Karnataka and Tamil Nadu.

Cinnamon is a medium–tall tree that reaches up to a height of 6–15 m on full growth. Cinnamon is one of the oldest–known spices that have been traded since time immemorial.

12.1. CLIMATE

Cinnamon is adapted to a wide range of climatic conditions. It grows well at an altitude of about 200 m from sea level. Annual rainfall of 200–250 cm and an average temperature of 27°C is ideal for its cultivation. Cinnamon is commercially cultivated as a rain fed crop. Prolonged spells of dry weather are not conducive for successful growth of Cinnamon.

12.2. SOIL

Cinnamon can be grown on a wide variety of soil. In the west coast of India, Cinnamon is grown on laterite soil and sandy patches with poor nutrient status.

12.3. PROPAGATION

Generally, seed propagation and vegetative propagation through cuttings and air layers are practiced in Cinnamon. Seeds are sown in sand beds or polythene bags containing a mixture of sand, soil and well–powdered dried cow dung in a 3:3:1 ratio.

Fig 26. Cinnamon

Pits of 50 cm are dug at a spacing of 3 x 3 m and pits are filled with compost and top soil before planting. Cinnamon is planted during the month of June–July.

12.4. MANURING

The manuring schedule for Cinnamon per plant is as follows:

1st year

Nitrogen (N)	: 20 g
Phosphorus (P_2O_5)	: 18 g
Potassium (K_2O)	: 25 g

3 YAP (Years After Planting)

Compost or FYM	: 29.0 kg
Neem cake	: 4.0 kg
Nitrogen (N)	: 150 g
Phosphorus (P_2O_5)	: 75 g
Potassium (K_2O)	: 150 g

The manures should be applied in May–June and the fertilizers should be applied in two equal split doses, in May–June and September–October.

12.5. INTERCULTURAL OPERATION

1. Weeding

The land should be kept free from weeds during early stages of growth of the plant. Weeding should be done regularly to keep the weeds under check.

2. Training and pruning

When seedlings become 2–3 years old, the shoot is cut back to a height of 30 cm from ground level to produce side shoots. The side shoots growing from the base are to encourage the growth of more side shoots till the whole plant assumes the shape of a low bush. Sometimes, stooling is done by slight mounding of soil to encourage shoots.

12.6. IRRIGATION

Irrigation should be given to the seedlings till they get established. Thereafter, irrigation should be given as per the requirement of the plant. Cinnamon is a rain fed crop.

12.7. HARVESTING

Ideal time for harvesting shoots is September–November. Cinnamon is ready for harvesting after 2–3 yr from planting when the plant reaches the height of 1.5–2 m with 3–4 shoots, and the bark turns brown in colour. The stems should be cut close to the ground when they are about 2 yr old, as straight as possible, 1.0–1.25 m length and 1.25 cm thickness. Harvested shoots should be bundled together and transported to the pack house for further post–harvest procedures.

12.8. YIELD

The yield varies with the age of the plant as follows:

Table 80. Yield of Cinnamon

Sr. No.	Age of the plant	Yield(quills/ha)
1.	3–4 yr and onwards	60.0–120.0 kg
2.	10–11 yr and onwards	225.0–300.0 kg

12.9. VARIETIES

Indian Varieties: Navashree, Nithyashree and Yercaud-1 Konkan Tej, Sugandini. Pani-Miris Kurundu, Sevel Kurundu, Tithta Kurundu, Vijaya, Sri Gamunu etc.

Sri Lankan varieties: Pat, Ma-pat, Kurunchi, Sweet Cimamon and Honey Cinnamon etc.

12.10. MEDICINAL VALUE

Cinnamon has some medicinal values as follows:

- Cinnamon helps in lowering the cholesterol, regulating the blood sugar and preventing the cancer.
- Cinnamon is an anti–clotting agent, anti–bacterial agent and also useful in getting relief from Arthritis.
- Cinnamon is useful in maintaining and improving brain health and works as an *E. coli* fighter.

12.11. PLANT PROTECTION

Insect pests

1. Gall–forming mites

Scientific name - *Erioplytes boisi* Gerber

Nature of damage

Mites lay eggs in the leaf tissue and gall formation starts on the lower surface of leaves immediately. The galls are pinkish in colour at the beginning and turn to green over maturity.

Control measures

- The disease–infested leaf should be collected and destroyed by burning.
- The insect can be controlled by spraying the crop with any one of the insecticides given in Table 81.

Table 81. Insecticides used for controlling the Gall forming mites of Cinnamon

Sr. No.	Name of insecticide	Amount (per litre of water)
1.	Acephate 75 WP	0.75 g
2.	Sulphur 80 WP	3.0 g
3.	Milbimectin 1 EC	0.5 ml
4.	Fenazaquin 10 EC	2 ml

2. Cinnamon butterfly

Scientific name - *Chilasa clytie*

Nature of damage

Larvae feed on tender and slightly mature leaves and causes defoliation in severe infestation. Only midribs of leaves with portions of veins are left behind.

Control measures

The insect can be controlled by spraying the crop with any one of the insecticides given in Table 82.

Table 82. Insecticides used for controlling the Cinnamon butterfly

Sr. No.	Name of insecticide	Amount (per litre of water)
1.	Acephate 75 WP	0.75 g
2.	Imidachloropid 17.8 SL	0.2 ml

3. Shoot and leaf webber

Scientific name - *Sorolopha archimedias*

Nature of damage

The larvae are very active and they web the leaves and the terminal shoots into clusters. The webbed cluster of leaves harbour several larvae and they are gregarious at the beginning and feed by scrapping the leaf surface.

Control measures

- Pruning the infested clusters is a good cultural control method.
- The insect can be controlled by spraying the crop with any one of the insecticides given in Table 83.

Table 83. Insecticides used for controlling the Shoot and leaf webber of Cinnamon

Sr. No.	Name of insecticide	Amount (per litre of water)
1.	Acephate 75 WP	0.75 g
2.	Thiomethoxam 25 WG	0.2 g
3.	Imidachloropid 17.8 SL	0.2 ml

4. Leaf miner

Scientific name - *Conopomorpha civica*

Nature of damage

Larvae feed on the tissues between the upper and lower epidermis of tender leaves resulting in linear mines that end in 'blister–like' patches. The infested leaves become crinkled and the mined areas dry up leading to the formation of large holes on the leaves.

Control measures

The insect can be controlled by spraying the crop with any one of the insecticides given in Table 84.

Table 84. Insecticides used for controlling the Leaf miner of Cinnamon

Sr. No.	Name of insecticide	Amount (per litre of water)
1.	Acephate 75 WP	0.75 g
2.	Thiomethoxam 25 WG	0.2 g
3.	Imidachloropid 17.8 SL	0.2 ml

5. Pink stem borer

Scientific name - *Ichneumoniptera cf. xanthosoma*

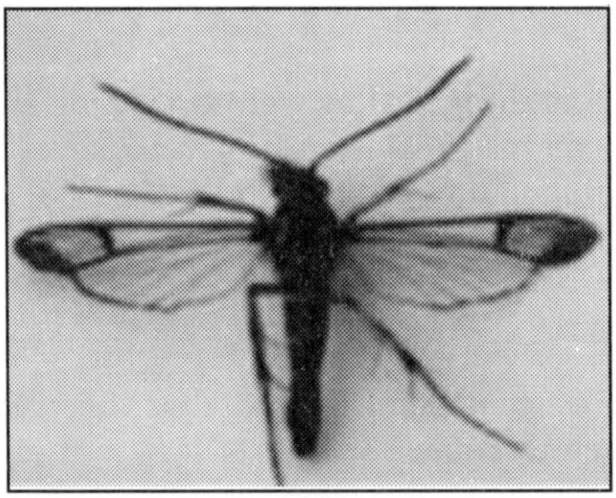

Fig 27. Pink stem borer (Adult)

Nature of damage

The larva eats into the plant stem near the soil surface. As a result, new shoots may die and some mature shoots may collapse from the base. New shoot formation is also retarded. At the end, gradually the bush dies. The insect causes damage to the old plantation.

Control measures

- The damage can successfully be controlled by covering the plant base by earthing up with soil.
- The insect can be controlled by spraying any one of the insecticides given in Table 85, on the base of the plant:

Table 85. Insecticides used for controlling the Pink stem borer of Cinnamon

Sr. No.	Name of insecticide	Amount (per litre of water)
1.	Carbosulfan 25 EC	2.0 ml
2.	Chloropyriphos 20 EC	2.5 ml
3.	Thiomethoxam 25 WG	0.2 g

Diseases

1. Rough bark disease

C.O.: *Phomopsis* sp

Rough bark disease is the most common disease of Cinnamon.

Fig 28. Rough bark disease of Cinnamon

Symptoms

Leaves of the infested plants show clorosis. Under severe conditions, infected immature plants die. The disease producing organisms affects the young bark of immature shoots by developing brown spots and spread gradually throughout the bark.

Control measures

- Disease can be controlled by destroying diseased plant and through adoption of correct cultural practices.
- The disease can be controlled by spraying the plant with any one of the fungicides given in Table 86.

Table 86. Fungicides used for controlling the Rough bark disease of Cinnamon

Sr. No.	Name of fungicide	Amount (per litre of water)
1.	Mancozeb 75 WP	2.5 g
2.	Carbendazim 50 WP	1.0 g
3.	Copper oxychloride 50 WP	5.0 g

2. White root disease

C.O.: *Fomes noxis*

Symptoms

Yellowing and subsequent shedding of leaves and sudden death of plants are visible symptoms. White coloured fungal mycelia growths can be observed on roots of infected plants.

Control measures

- The diseased–dead plants should be uprooted and destroyed by burning.
- The root bases should be cleaned.

Sulphur powder should be applied to the bases of infected plants and planting holes when new plants are establishing in such lands.

3. Leaf spot or Leaf blight

C.O.: *Colletotrichum gloeosporides*

Symptoms

Symptoms are seen as brownish leaf spots and these may enlarge to make large lesion.

Control measures

The disease can be controlled by spraying the plant with any one of the fungicides given in Table 87.

Table 87. Fungicides used for controlling the Leaf spot or Leaf blight of Cinnamon

Sr. No.	Name of fungicide	Amount (per litre of water)
1.	Mancozeb 75 WP	2.5 g
2.	Carbendazim 50 WP	1.0 g
3.	Copper oxychloride 50 WP	5.0 g
4.	Thiophenate methyl 70 WP	1.0 g
5.	Metalaxyl 35 WS	0.5 g

4. Black sooty mold

C.O.: *Stenella* sp

Symptoms

The blackish growth on the leaf surface is the characteristic symptom of the disease. The fungal growth is confined only to the surface and there is no penetration into leaf tissues.

Control measures

The disease does not affect the yield severely. So, application of fungicides is not necessary.

13 GARLIC

Botanical name - *Allium sativum*

Family - Liliaceae

Commercial part - Bulb

Indian names - *Bengali* : Rasun; *Assamese* : Naharu; *Hindi* : Lasun, Lessan, Lahsun; *Gujarati* : Lasan; *Kannada* : Bellulli; *Kashmiri* : Ruhan; *Malayalam* : Vellulli; *Marathi* : Lusson; *Oriya* : Rasuna; *Punjabi* : Lassan, Lasun; *Sanskrit* : Lashuna; *Tamil* : Ullipundu, Vellaippundu; *Telugu* : Velluri; *Urdu* : Lassun, Leshun.

Garlic is one of the most important bulb crops. It is used as a spice or condiment throughout India. The compound bulb of Garlic consists of small bulblets or cloves.

13.1. ORIGIN AND DISTRIBUTION

Garlic is a native of west Asia and Mediterranean area. China, Korea, India, USA, Spain, Argentina and Egypt are the major Garlic–growing countries.

Fresh peeled Garlic cloves have the following composition:

Table 88. Nutrient in Garlic (100 g)

Moisture	62.0 g
Carbohydrates	29.0 g
Protein	0.001 mg
Total ash	310 mg
Fibre	30 mg
Fat	0.1 g
Calcium	1.0 g

Phosphorus	0.8 g
Iron	6.3 g
Copper	2.5 mg
Vitamin B-12	0.6 mg
Vitamin B-6	11 mg
Vitamin C	13 mg
Vitamin B-9	3 mg
Nicotinic acid	0.4 mg
Tryptophan	3.1 mg
Selenium	3.6 mg
Manganese	15 mg
Calorific value	142

13.2. CLIMATE

Garlic is a cool season crop and it succeeds best in mild season without extreme of heat and cold. The suitable growth temperature for Garlic is 13°C–24°C. Short days are favourable for bulb formation. Garlic is a frost tolerant crop. It requires cool and moist climate during growth period and warm dry weather during maturity.

Fig. 29. Garlic cloves

13.3. SOIL

Garlic can be grown on a variety of soils but it thrives better on fertile, well–drained loam, sandy loam and silty loam soils. The pH of soil between 6.0 and 7.0 is suitable for a good crop. Highly alkaline and saline soils are not suitable for Garlic cultivation.

13.4. LAND PREPARATION

Land should be brought to fine tilth by shallow ploughing 4–5 times with tractor or powertiller giving sufficient intervals between two ploughings. Land should be levelled and divided into small plots and channels to facilitate proper irrigation and intercultural operations. The manures should be applied at the time of land preparation and mixed well with soil by ploughing followed by planking.

13.5. PLANTING

Garlic is propagated by cloves. Well–grown compact bulbs of uniform shape and size should be selected. The cloves having 8–10 cm size should be used for planting.

1. Time of planting

Table 89. Time of planting - Garlic

S.N.	Name of State	Sowing time
1.	Madhya Pradesh, Maharashtra, Karnataka and Andhra Pradesh	August–October
2.	Gujarat	October–November
3.	West Bengal	October–November

Proper season for planting in higher hills in northern part is March–April. It can also be, however, planted during September–October. Garlic is planted from September–November in northern plains of India

2. Seed rate

About 500 kg cloves of 8–10 mm diameter are required for planting in one hectare.

3. Seed treatment

The bulbs suspected of being diseased should be treated with registered chemicals given in Table 90, prior to planting.

Table 90. Fungicides for seed treatment of Garlic

Sr. No.	Name of fungicide/bio–fungicide	Amount (per litre of water)
1.	Carbendazim 50 WP	2.0–3.0 g
2.	M.E.M.C 6 WP	1.0 g
3.	Captan 50 WP	2.0–5.0 g
4.	Mancozeb 75 WP	2.5 g
5.	Trichoderma viridi	4.0–5.0 g

4. Method of planting

Garlic is propagated from cloves or bulblets. Cloves of 8–10 mm should be taken to select bigger cloves from outer side of bulbs. Healthy cloves or bulblets free from diseases and injuries should be used for planting.

- **Dibbling method** - The Garlic is generally planted by dibbling method at 5.0–7.5 cm deep, keeping their growing end upward. Then they should be covered with loose soil.
- **Furrow planting** - The furrows should be made with small country plough or by spade. The cloves should be placed in furrows and covered with loose soil.

Spacing

Row to row: 15 cm

Plant to plant: 7.5 cm

13.6. MANURING

The recommended fertilizer dose for Garlic is as follows:

1. Basal dressing

Manures

Compost or FYM	: 10.0–15.0 t/ha
Azophos	: 15.0 kg/ha

Fertilizers

Nitrogen (N)	: 50.0 kg/ha
Phosphorus (P_2O_5)	: 50.0 kg/ha
Potassium (K_2O)	: 50.0 kg/ha

2. Top dressing

Nitrogen(N)

- 25.0 kg/ha at 25 DAP
- 25.0 kg/ha at 42 DAP

Micronutrients

Micronutrients also increase the yield potential. The amount of micronutrients to be used is as follows:

- Zinc (Zn)

  Zinc Sulphate: 25.0 kg/ha
- Boron (B)

  Borax: 10.0 kg/ha
- Molybdenum (Mo)

  Ammonium molybdate: 0.5 kg/ha

Micronutrients should be applied once in 3–4 crops on the basis of soil testing report.

13.7. IRRIGATION

Garlic being a shallow–rooted crop, needs frequent irrigation. There should be no moisture scarcity during the growth period. In general, Garlic needs irrigation at eight days intervals during vegetative growth and 10–15 d during maturation. Chauhan (1972) stated that the first irrigation should be given after sowing and then the field should be irrigated after every 10–15 d till the season warms up. There should be no scarcity of moisture in the growing season; otherwise the development of bulbs will be hampered. When the bulbs mature, irrigation should be done sparingly. The last irrigation should be given 2–3 d before harvesting to make the harvesting easy without damaging the bulbs. Irrigation after a long spell of drought results in splitting of bulbs. Excessive irrigation results in sprouting.

13.8. INTERCULTURAL OPERATION

First and second weeding should be given with hand hoe or khurpi, one and two months after planting of the crop respectively. According to Joshi (1961), hoeing the crop just before the formation of bulbs (about 2 ½ months from sowing) loosens the soil and helps in setting of bigger and well–filled bulbs. The crop should not be weeded out or hoed at a later stage because this may damage the stem and impair the keeping quality of sprouting cloves and the cloves already formed.

The weeds may also be controlled by application of herbicides as follows:

Table 91. List of herbicides for controlling weeds in Garlic

Sr. No.	Name of herbicide	Time of application	Amount (per hectare)
1.	Glyphosate 41 SL	Post–emergence	800–2,000 g
2.	Pendimethalin 30 Ec	Pre–emergence	1.5–2.5 ml

Application of herbicides plus one hand weeding gives good control over broad–leaved weeds.

13.9. HARVESTING AND CURING

Garlic is a crop of four and half to five month's duration. The ideal timing of harvest is when the bulbs attain maximum size but the cloves have not started to separate. A good index of maturity is when the main stem above the bulb softens, the top has fallen, and when 75 percent of the leaves turn yellow.

Garlic becomes ready for harvesting when its tops turn yellowish or brownish in colour, and show signs of drying up and bending over. The plants should be pulled or uprooted with spade or khurpi. The harvested plants should be tied into small bundles and kept under shade for 2–3 d for curing.

13.10. VARIETIES

Godavari, Sweta, HG-1, HG-6, Pusa Sel-10, LCG-1, ARU-52, Agrifound White (G-41), Yamuna Safed (G-1), Yamuna Safed-2 (G-50), G-282, Agrifound Parvati, Agri-pride, Solan, Pusa mil, Fawari, Rajli gaddi, G-41, Selection- 2, Selection-10, Garlic 9 etc.

13.11. YIELD

The yield of the bulbs varies from 100 to 200 q/ha depending upon variety and regions.

13.12. USES

Garlic has many health benefits as follows:

- Garlic boosts immunity and wards off colds and coughs, including whooping cough.
- Garlic is used for removing skin infections like athlete's foot, abrasions and acne.
- Garlic is invaluable for lowering blood pressure as well as preventing cardiac problems. It also prevents degenerative conditions related to diabetes such as kidney problems and retinopathy. Moreover, it detoxifies the liver, boosts digestion and prevents acidity.
- Garlic also helps to expel worms.
- It is said that Garlic improves memory.

13.13. PLANT PROTECTION

Insect pests

1. Thrips

Scientific name - *Thrips* sp

The thrip is most serious insect pests of Garlic.

Nature of damage

Nymphs and adults suck plant sap, especially from the young leaves and growing points. The leaves turn yellow, curl slightly and become silvery–white or brown. Small black dots become visible. If the infestation is severe, the leaves wilt and die.

Control measures

- Nymphs usually hide in the angles of leaves, making it difficult to control them. So, the badly–infested leaves and bulbs should be hand picked and destroyed by burning or burrowing under the soil.
- Early planting is recommended in October to avoid heavy infestation, which usually occurs in late January to March.
- The insect can be controlled by using insecticides, starting from when the plants are 30–60 days old. Frequency of application can be 5–10 days intervals depending on degree of infestation. The thrips can be controlled by spraying any one of the insecticides given in Table 92.

Table 92. Insecticide used for controlling the Thrips of Garlic

Sr. No.	Name of insecticide	Amount (per litre of water)
1.	Acephate 75 WP	0.75 g
2.	Quinalphos 25 EC	2.0 ml
3.	Clothianidin 50 WDG	0.1 gm
4.	Triazophos 40 EC	2.0 ml

2. Mites

Scientific name - *Aceria tulipae*

The mites are so small that their presence in plants is often unnoticed. They are either seed–borne or mulch–borne. Mite damage to Garlic is often called 'tangle top'.

Nature of damage

The mites suck the sap from the plants and cause serious damage to the crop. Affected plants become twisted and distorted with yellowish or pale green–coloured streaks on the leaves. The leaves may also not emerge readily from the cloves; they separate poorly after emergence.

Control measures

- Twisted and unextruded young leaves can also be stretched, but this should be done in the afternoon when the leaves are somewhat wilted to avoid breakage.
- The insect can be controlled by spraying any one of the insecticides given in Table 93, every 7–14 days, depending on extent of infestation:

Table 93. Insecticides used for controlling the Mites of Garlic

Sr. No.	Name of insecticide	Amount (per litre of water)
1.	Dicofol 18.5 EC	1.5 ml
2.	Sulphur 80 WP	3.0g
3.	Milbimectin 1 EC	0.5 ml
4.	Properjit 57EC	2.0 ml

3. Tobacco moth

Scientific name - *Ephestia elutella*

The tobacco moth is the most serious storage insect pest of Garlic.

Nature of damage

The larvae enter into individual cloves of bulbs and feed on the contents, leaving the cloves empty and dry. In serious cases, the damaged cloves turn powdery.

Control measures

- Bulbs selected for seed purposes should be treated with insecticides before storing. These insecticide–treated bulbs should be air–dried for one day, and then they should be stored at room temperature or bundled and hanged in a well–ventilated place.
- The insect can be controlled by spraying the crop with any one of the insecticides given in Table 94.

Table 94. Insecticides used for controlling the Tobacco moth of Garlic

Sr. No.	Name of insecticide	Amount (per litre of water)
1.	Acephate 75 WP	1.0 g
2.	Thiomethoxcam 25 WG	0.2 g
3.	Imidachloropid 17.8 SL	0.2 ml
4.	Chlorpyriphos 20 EC	2.5 ml

4. Pink stalk borer

Scientific name - *Sesamia calamistis*

The larvae of the moth attack the plant and cause damage to the crop.

Nature of damage

The larvae bore into the stems of the Garlic plants. This causes yellowing of the leaf tips and eventually leading to the die off of the central leaves.

Control measures

- Pink stalk borer can be reduced by cultural practices such as weed control and removal of alternative host plants in the vicinity of Garlic fields.
- Natural enemies of the pink stalk borer include the parasitic wasps of the families *Braconidae and Ichneumonidae*, which can be used for the control of this insect pest biologically.
- The insect can be controlled by spraying the crop with any one of the insecticides given in Table 95.

Table 95. Insecticides used for controlling the Pink stalk borer of Garlic

Sr. No.	Name of insecticide	Amount (per litre of water)
1.	Acephate 75 WP	1.0 g
2.	Thiomethoxam 25 WG	0.2 g
3.	Imidachloropid 17.8 SL	0.2 ml
4.	Chlorpyriphos 20 EC	2.5 ml
5.	Neem oil 1,500 ppm	3.0 ml

5. Garlic maggot

Scientific name - *Pegomya cepetorium* Meade

Garlic maggot also causes the damage to Garlic.

Nature of damage

The Garlic maggot is the offspring of a small fly that lays eggs near the base of the plant or on the bulb itself. The maggots kill the plant by burrowing into the stem and the bulb.

Control measures

- The plants that are affected by the maggot should be destroyed before the maggots mature into flies.

- Application consists of placing a cupful of kerosene oil in a bucket of dry sand at the base along the row. It is effective for deterring the parent flies from depositing their eggs. This also kills the young maggots that might attempt to work through the sand (Shanmugavelu 1989).

6. Black garlic fly

Scientific name - *Tritora flexa*

The Black garlic fly is mentioned as an occasional insect pest of Onions and Garlic that can cause some damage to the bulbs, but has never been elevated to 'serious pest' status.

Nature of damage

The Black garlic fly feeds on Garlic and chews the Garlic both in the field and in storage.

Control measures

- Collection and destruction of all infested parts of the Garlic help in minimizing the damage to the crop.
- Dusting the crop with Malathion 5D or Carbaryl 50 WP @ 25.0 kg/ha in the insect–affected field helps in controlling the Black garlic fly.
- Use of poison bait traps prepared from different ingredients, as follows, is most effective in controlling this insect pest.

Mollases	: 200 g
Malathion 50 EC	: 3 ml
Water	: 2 l

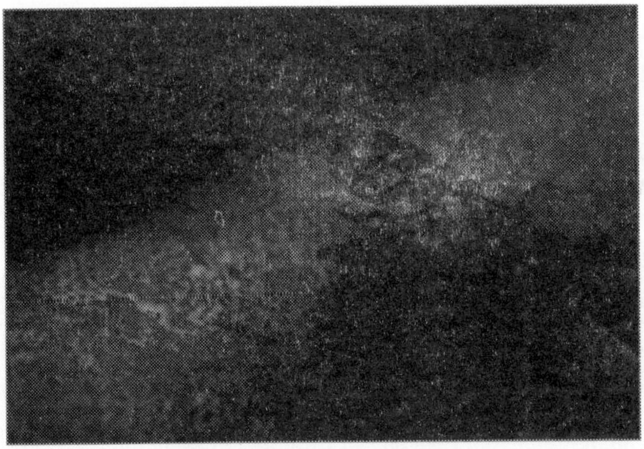

Fig 30. Black garlic fly

The poison bait traps should be taken in small earthen pots and should be kept in different places in the field.

- Spraying the crop with any one of the insecticides given in Table 95 is found to being effective in controlling the Black onion fly.

Table 96. Insecticides used for controlling the Black garlic fly

Sr. No.	Name of insecticide	Amount (per litre of water)
1.	Profenophos 50 EC	1.5 ml
2.	Acephate 75WP	0.75 g
3.	Neem oil 1,500 ppm	3.0 ml
4.	Imidachloropid 17.8 EC	0.2 ml

Diseases

1. White bulb rot of Garlic

C.O.: *Sclerotium cepivorus*

The white bulb rot of Garlic is a common disease in Garlic. The disease is more destructive during the early spring and the autumn seasons.

Symptoms

This disease attacks the crop at all the stages of growth in the field, and the leaves of both partly and fully–developed plants turn yellow and wilt. The snow–white mycelia are seen on the bulb surface or at the crown of the seedlings. The root system of the crops should be destroyed completely. The disease also attacks cloves of Garlic in storage.

Control measures

In field

- Crop rotation and thorough destruction of infected material is the best control measure of the disease.
- The disease can be controlled by spraying any one of the fungicides given in Table 97.

Table 97. Fungicides use for controlling white bulb rot of Garlic

Sr. No.	Name of fungicide	Amount (per litre of water)
1.	Mancozeb 75 WP	2.5 g
2.	Copper oxychloride 50 WP	5.0 g
3.	Tridemorph 80 EC	0.5 ml
4.	Carbendazim 50 WP	1.0 g

In storage

- The infected bulb should be sorted out and destroyed by burrowing in the soil.
- The bulbs that are stored in the storage should be turned frequently to check the spread of the disease.

2. Brown rust

C.O.: *Puccinia alli*

Symptoms

The rust pustules, which are bright orange in colour, slightly raised and elongated, appear on the leaves. The leaf is also covered with powdery spores, and turns to a yellow colour as time goes on. This leaf disease can be responsible for serious losses under prolonged moist conditions.

Control measures

The disease can be controlled by spraying any one of the fungicides given in Table 98.

Table 98. Fungicides used for controlling the Brown rust of Garlic

Sr. No.	Name of fungicide	Amount (per litre of water)
1.	Tridemorph 80 EC	0.5 ml
2.	Carbendazim 50 WP	1.0 g
3.	Metalaxyl 35 WS	0.5 g
4.	Flutriazole 5 SC	2.0 g

3. Pink root

C.O.: *Pyrenochaeta terrestris*

The pink rot disease can survive on roots of different vegetable crops.

Symptoms

The infected crop becomes poorly defined and the plants become stunted as compared to the other healthy plants. The roots become pink and rot. The length and the number of leaves are also reduced. This disease also results in smaller bulbs that will be harvested.

Control measures

- The bulbs from infected fields should not be used for planting. The field of production that is infected can be flooded with water to control this disease.

- The bulbs suspected of being diseased should be treated with a registered chemical prior to planting.
- The Pink rot disease can be controlled by spraying any one of the fungicides given in Table 99.

Table 99. Fungicides used for controlling the Pink root of Garlic

Sr. No.	Name of fungicide/bio-fungicide	Amount (per litre of water)
1.	Carbendazim 50 WP	2.0–3.0 g
2.	M.E.M.C 6 WP	1.0 g
3.	Captan 50 WP	2.0–5.0 g
4.	Mancozeb 75 WP	2.5 g
5.	Trichoderma viridi	4.0–5.0 g

- A sound crop–rotation system should be followed.

4. Fusarium bulb rot

C.O.: *Fusarium oxyporum* F *cepae* (Hanzwa), Snyder and Jausen.

Symptoms

The disease–producing organism attacks the basal part of Garlic, especially roots and bulbs, which become semi–watery and rotten. This causes progressive yellowing and drying of leaves. Eventually, the top parts of the plants collapse.

Control measures

- The disease–free bulbs, collected from healthy crop, should be used for planting.
- All field–refuse after harvesting should be collected and destroyed by burning. Summer ploughing or ploughing the land and allowing it to dry for at least three months before planting is found good to reduce the disease incidence.
- The crop rotation should be practiced.
- The disease can be controlled by spraying the crop with any one of the fungicides given in Table 100.

Table 100. Fungicides used for controlling the Fusarium bulb rot of Garlic

Sr. No.	Name of fungicide	Amount (per litre of water)
1.	Carbendazim 50 WP	1.0 g
2.	Copper oxychloride 50 WP	5.0 g
3.	Captan 50 WP	2.5 g
4.	Mancozeb 75 WP	2.5 g

5. Cercospora leaf spot

C.O.: *Cercospora duddiae*

Symptoms

The disease appears on leaves as small, ash–coloured and irregular–shaped spots, about 3–5 mm in diameter scattered on leaf lamina. The spots coalesce gradually and results in blightening of foliage. The disease progresses rapidly after the original lesion fuses. Affected leaves eventually die.

Control measures

- Clean cultivation such as practice of crop rotation, removal and burning of weeds and maintaining good field drainage helps in reducing the disease incidence.
- The disease–free planting materials, collected from healthy crop, should be used.
- The disease can be controlled by spraying the crop with any of the fungicides given in Table 101 at 7–10 days interval:

Table 101. Fungicides used for controlling the Cercospora leaf rot of Garlic

Sr. No.	Name of fungicide	Amount (per litre of water)
1.	Copper oxychloride 50WP	5.0 g
2.	Captan 50WP	2.5 g
3.	Ziram 27 SL	3.0 ml
4.	Thiophenate methyl 70 WP	1.0 g

6. Purple blotch

C.O.: *Alternaria porri* (Ellis)

Symptoms

Purple blotch appears on leaves as small whitish sunken lesions with purple centres that rapidly enlarge. The margin of the lesion is often a shade of red or purple and is surrounded by a yellow zone. Infected leaves turn yellow and soon dry up completely and fall over gradually. Seed stalks, flowers of seed onions and, at harvest, bulbs may become infected.

Control measures

- The disease–free planting materials, collected from healthy crop should be used.

- The disease can be controlled by spraying the crop with any of the fungicides given in Table 102 at 7–10 d intervals:

Table 102. Fungicides used for controlling the Purple blotch of Garlic

Sr. No.	Name of fungicide	Amount (per litre of water)
1.	Copper oxychloride 50 WP	5.0 g
2.	Mancozeb 75 WP	2.5 g
3.	Carbendazim 50 WP	1.0 g
4.	Zineb 75 WP	2.5 g

7. Downy mildew

C.O.: *Peronospora destructor*

Symptoms

Elongated patches up to 1.25 inch long appear on the leaf that are somewhat pale than the rest of the foliage or have turned a light tan to brown in colour. A furry growth, greyish violet in colour, may be visible on the surface of the infected leaves or seeds or stems during moist periods, especially early morning. Downy mildew lesions may be violet to purple in colour. Infected leaves gradually turn pale green to yellow in colour and diseased parts such as leaf tips fold over and collapse.

Control measures

Preventive applications of any one of the fungicides in Table 103 to the foliage, when climatic conditions favour disease development, can effectively control downy mildew:

Table 103. Fungicides used for controlling the Downy mildew of Garlic

Sr. No.	Name of fungicide	Amount (per litre of water)
1.	Mancozeb 75 WP	2.5 g
2.	Chlorothalonil 75 WP	2.0 g
3.	Copper hydroxide 77 WP	2.0 g

Two or more of these materials should be rotated throughout the season as a resistance management strategy.

8. Powdery mildew

C.O.: *Oidiopsis* sp

Symptoms

Powdery mildew on Garlic causes light yellow lesions on leaves within which single spores (conidia) on clusters of stalks (conidiophores) emerge from stomatal openings. Distinct pale-yellow patches of variable size on abaxial surface of leaf associated with powdery mass are its main characteristics.

Control measures

- The crop residues should be destroyed as the fungus perpetuates in the debris of the plant.
- Late planting should be avoided.
- The disease can be controlled by dusting or spraying any of the fungicides given in Table 104 on the appearance of the disease at an interval of 7–10 days.

Table 104. Fungicides used for controlling the Powdery mildew of Garlic

Sr. No.	Name of fungicide	Amount (per litre of water)
1.	Sulphur 80 WP	3.0 g
2.	Dithiolone 75 WP	2.5 g
3.	Carbendazim 50 WP	1.0 g

14 ONION

Botanical name - *Allium cepa* Linn.

Family - Liliaceae

Commercial part - Bulb

Indian names - *Bengali* : Piyaz; *Hindi* : Piyaz; *Punjabi* : Piyaz; *Urdu* : Piyaz; *Assamese* : Piyaz; *Gujarati* : Dunzari; *Kannada* : Nirulli; *Malayalam* : Bawa; *Marathi* : Kanda; *Oriya* : Piyaza; *Sanskrit* : Palandu; *Sindhi* : Dungari; *Tamil* : Gayam, Irulli; *Telugu* : Nirulli.

Onion (*Allium cepa*), which is also known as the Bulb onion, is the most widely–cultivated spice of the genus Allium. The Onion is one of the most important commercial vegetable crops grown in India. Maharashtra, Tamil Nadu, Andhra Pradesh, Bihar and Punjab are the important Onion–producing States.

Table 105. Nutritive value of Onion

Raw Onions	
Nutritional value per 100 g (3.5 oz)	
Energy	166 kJ (40 kcal)
Carbohydrates	9.34 g
- Sugars	4.24 g
- Dietary fibre	1.7 g
Fat	0.1 g
- Saturated	0.042 g
- Monounsaturated	0.013 g
- Polyunsaturated	0.017 g
Protein	1.1 g

Water	89.11 g
Vitamin A equiv.	0 µg (0%)
Thiamine (vit. B_1)	0.046 mg (4%)
Riboflavin (vit. B_2)	0.027 mg (2%)
Niacin (vit. B_3)	0.116 mg (1%)
Vitamin B_6	0.12 mg (9%)
Folate (vit. B_9)	19 µg (5%)
Vitamin B_{12}	0 µg (0%)
Vitamin C	7.4 mg (9%)
Vitamin E	0.02 mg (0%)
Vitamin K	0.4 µg (0%)
Calcium	23 mg (2%)
Iron	0.21 mg (2%)
Magnesium	0.129 mg (0%)
Phosphorus	29 mg (4%)
Potassium	146 mg (3%)
Sodium	4 mg (0%)
Zinc	0.17 mg (2%)

Percentages are relative to US recommendations for adults.
Source: USDA Nutrient Database

14.1. CLIMATE

Onion is a cool season crop. It is grown during winter and harvested before the real summer season begins. Onion can be grown under a wide range of climatic conditions. It requires different range of temperatures at different growth stages as follows:

Vegetative growth	: 15°C–21°C
Bulb formation	: 20°C–25°C
Seedling growth	: 20°C–25°C

It succeeds best in mild season without extremes of heat and cold. According to Rao and Purewal (1954), it can be grown as a rainfed crop even at an elevation of 1,500 to 2,000 m between April and August. Onion produces bulbs more rapidly at warm temperature (21°C–26°C) than at cool temperature (15°C–21°C) (Thompson and Smith 1938). The growth starts declining at temperatures higher than 27°C.

14.2. SOIL

Onion can be grown in all types of soils. But, sandy loam, silty loam and deep friable soil are best suited for Onion crop. Heavy soil (i.e., clay soil) is not suitable for the cultivation of Onion as heavy soils give rise to ill–shaped bulbs causing problem in digging of bulbs. The optimum pH range of the soil needs to be between 5.8 and 6.5 (Chaudhury 1967). Highly alkaline and saline soils are not suitable for Onion cultivation. Good drainage is essential for this crop. Waterlogging can result in failure of the crop.

14.3. LAND PREPARATION

Land should be brought to fine tilth by shallow ploughing 4-5 times with tractor or power tiller giving sufficient intervals between two ploughings. Land should be levelled and divided into small plots and channels to facilitate proper irrigation and intercultural operations. The manures should be applied at the time of land preparation and mixed well with the soil by ploughing followed by planking.

14.4. PROPAGATION

Onion is propagated by planting seedlings and bulbs.

1. Raising seedlings

Onion seeds should be sown in a raised nursery bed, about 3 m long, 0.6 m wide and 20–30 cm high to raise seedlings. About 70 cm distance should be kept between two beds to carry out operations of watering, weeding etc. The surface of beds should be smooth and well–levelled. Before sowing, seed should be treated with any one of the fungicides given in Table 106.

Table 106. Fungicides used for seed treatment of Onion

Sr. No.	Names of fungicides/bio-fungicides	Amount (per litre of water)
1.	Carbendazim 50 WP	2.0–3.0 g
2.	M.E.M.C 6 WP	1.0 g
3.	Captan 50 WP	2.0–5.0 g
4.	Mancozeb 75 WP	2.5 g
5.	Trichoderma viridi	4.0–5.0 g

The soil in the nursery should also be treated with Captan 75 SD @ 4–5 g/m^2 area. The nursery beds should be irrigated 15–20 d before sowing and should be covered with transparent polythene of 250 gauge for soil solarisation. Application of Trichoderma viride @ 1,250 g/ha is also recommended to manage damping off and to raise healthy seedlings. After sowing, the seeds should be covered with fine powdered FYM or compost followed by light watering. Then, the beds should be covered with dry straw or grass or sugarcane leaves to maintain optimum temperature and moisture, which should be removed immediately after germination is completed.

Seed rate: 8.0–10.0 kg/ha

The seedlings become ready for transplanting in about eight weeks when they are about 6 inches (15 cm) tall (Singh 1967). According to Dr. Mehta (1959), it is better to plant medium seedlings than the smaller or taller ones. Rao et al 1967) advised transplanting of seedlings at a spacing of 7.5 x 10 cm.

2. Sowing seeds directly in the main field

The seeds should be sown directly in the field by broadcasting or drilling in rows 30 cm apart during August–September in plains and February–June in hilly areas. The plants may be thinned to proper distance when they are 6–8 weeks old.

Seed rate : 20.0–25.0 kg/ha

3. Planting bulb in the field

The bulb or bulbils are planted in the main field for early green crop of Onion. They should be planted on ridges in case of early planting or in beds or furrows prepared by country plough or by using spade at 15 cm apart in the rows of 45 cm.

Seed rate: 10.0-12.0 q/ha

14.5. MANURING

Manures

Compost or FYM : 10.0 t/ha

Azophos : 15.0 kg/ha

Fertilizers

Nitrogen (N) : 150.0 kg/ha

Phosphorus (P_2O_5) : 100.0 kg/ha

Potassium (K_2O) : 150.0 kg/ha

Micronutrients

Zinc (Zn)

Zinc sulphate : 25.0 kg/ha

Boron (B)

Borax : 10.0 kg/ha

Molybdenum (Mo)

Ammonium molybdate : 0.5 kg/ha

Method of applying fertilizers

Manures

Manures should be applied at the time of land preparation and mixed well in the soil by cross ploughing followed by planking.

Fertilizers

1. Basal dressing - Total amount of phosphorus, half amount of nitrogen and potassium.

2. Top dressing - One fourth amount of nitrogen and potassium at 21 DAT (Days after transplanting of seedling) / DAP (Days after planting of bulb) and rest amount at 40 DAT/DAP.

Micronutrients

The micronutrients should be applied at the time of land preparation if deficiency is observed in the soil test or in the previous crop. Micronutrients should be applied once in 3–4 crops.

14.6. SEASON

Onion is cultivated in two seasons as follows:

Rabi season - September–November

Kharif season - January–February

14.7. HARVESTING

Onion is harvested in different stages according to its uses as follows:

- **Vegetable purposes** - Onions for use in the green stage should be harvested as soon as they reach to edible size.
- **Mature stage** - The bulb crop becomes ready for harvest in 3–5 months after planting the bulb or transplanting the seedlings depending on the variety. Maturity is indicated by the top falling over, while the leaves are still green (Chauhan 1967). The Onion should be harvested when the tops have fallen over.

The lifting must be done very carefully because any wound to the bulbs gives ready access to ever present decay organisms.

The Onions should be placed on the ground in windrows with the tops covering the bulbs to prevent Sunscald. The Onions should be left in windrows until the tops become dry. The length of time required for the tops to dry depends on the weather and may be anywhere from 3–10 d.

After the tops are fairly well dried down, they should be cut off. About 1 inch of the top attached to the bulb should be left out. If the top is cut close, decay organisms have easy access to the bulb. Onions that show any signs of decay or have mechanical wounds or have thick necks should be discarded when the bulbs are being topped.

Curing

Onions that are to be stored for winter–use must be thoroughly cured. Curing process takes considerable time i.e., 3–4 weeks. The Onions should be cured in a place that is well ventilated such as a corn crib. The onions can be spread out on the floor where the air can circulate around them.

Another method that works well for the home gardener is to place the Onions in mesh bags (Onion sacks) and tie the sack to the rafters of a building that has free movement of air. A carport–type of building would be very good.

14.8. VARIETIES

Extant varieties before 1986

N-2-4-1, N-53, N-257-9-1, Pusa Red, Pusa White Flat, Pusa White Round, Pusa Ratnar, S-131, Early Grano, Punjab Selection, Punjab Red Round, Punjab-48, CO-1 (Multiplier), CO-2, CO-3, CO-4, MDU-1, Hissar-2, Kalyanpur Red Round, Udaipur 101, Brown Spanish (Long day) and VL-1 (Long day), and S-48

Fig 31. Onion (var. Arka Niketan)

Improved varieties after 1986:

Bhima Red, Bhima Raj, Bhima Super, Bhima Kiran, Bhima Sakti, Bhima Shubra, Bhima Shweta, Baswant-780, Arka Niketan, Arka Kalyan, Phule Safed, Phule Suwarna, Phule Samnarth, Agrifound Dark Red, Agrifound Light Red, Agrifound White, L-28, Akola Safed, Pusa Madhavi, Arka Bindu, Arka Pragati, Arka Pitambar, Arka Lalima (F1 hybrid), Arka Kirtiman (F1 hybrid), HOS-1, Agrifound Rose, Agrifound Red (Multiplier), Line-355, Udaipur 102, Udaipur 103, Punjab Naroya, Punjab White, VL-3 (Long day), Rajasthan Onion-1, and Aprita (RO-59).

Fig. 32. Onion (Var. Bhima Shewata)

Local/ farmers varieties

Fursungi Local (Pune), Sukhsagar (West Bengal), Bellary Red (Karnataka), K. P. Onion (Andhra Pradesh), Nasik Red (Nasik), Pillipatti Junagarh (Gujarat), Telgi Local (Vijapur) and Nimar Local (Madhya Pradesh).

Source: Directorate of Onion and Garlic Research

14.9. PLANT PROTECTION

Insect pests

1. Onion maggots

Scientific name - *Delia antique, D. platura*

Onion maggot is a common and destructive insect pest in Onion–producing regions of our country.

Nature of damage

The Onion maggot is the offspring of a small fly that lays eggs near the base of the plant or on the bulb itself. The maggots kill the plant by burrowing into the stem and the bulb.

Control measures

- The plants that are affected by the maggot should be destroyed before the maggots mature into flies.
- Application consists of placing sand soaked with a cupful of kerosene oil in a bucket of dry sand at the base along the row is effective for deterring the parent flies from depositing their eggs. This also kills the young maggots that might attempt to work through the sand (Shanmugavelu, 1989).

2. Thrips

Scientific name - *Thrips tabaci*

Nature of damage

Nymphs and adults suck plant sap, especially the young leaves and growing points. The leaves turn yellow, curl slightly and become silvery–white or brown in colour. Small black dots become visible on the leaves and if the infestation is severe, the leaves wilt and die.

Control measures

- Nymphs usually hide in the angles of leaves, rendering their control difficult. So the badly–infested leaves and bulbs should be handpicked and destroyed by burning or burrowing under the soil.
- Early planting is recommended in October to avoid heavy infestation, which usually occurs in late January–March.
- The insect attacks can be controlled by spraying any one of the insecticides given in Table 107, starting when the plants are 30–60 days old. Frequency of application should be 5–10 days intervals, depending on the degree of infestation. The thrips can be controlled by spraying any one of the insecticides given in Table 107.

Table 107. Insecticides used for controlling the Thrips of Onion

Sr. No.	Name of insecticide	Amount (per litre of water)
1.	Acephate 75 WP	0.75 g
2.	Quinalphos 25 EC	2.0 ml
3.	Clothianidin 50 WDG	0.1 g
4.	Triazophos 40 EC	2.0 ml

3. Onion fly

Scientific name - *Delia (Hylemya) antiqua*

The black onion fly is mentioned as an occasional pest of Onions and garlic that can cause some damage to the bulbs, but has never been elevated to 'serious pest' status.

Nature of damage

The black onion fly feeds on Onion and chews both in field and in stores. The first sign is the wilting of foliage; subsequently foliage becomes flaccid and collapses.

Control measures

- Collection and destruction of all infested parts of the Onion can help in minimizing the damage to the crop.
- Dusting the crop with Malathion 5D or Carbaryl 50WP @ 25.0 kg per hectare in the insect–affected field can help in controlling the black onion fly.
- Use of poison bait traps prepared from different ingredients, as follows, is most effective in controlling this insect pest:

Molasses	: 200 g
Malathion 50 EC	: 3 ml
Water	: 2 l

4. Onion leaf–eating caterpillar

Scientific name - *Spodoptera exigua*

Nature of damage

Larvae are defoliators. Young larvae feed gregariously and skeletonise the foliage. Older larvae are solitary and eat to make large holes in leaf tissue.

Control measures

- Bacillus thuringiensis (BT) is recommended @ 1.5-2.0 g per litre of water to control the caterpillar.
- The hand picking of caterpillar along with infected leaves at initial stages is found good for controlling the insect pest.
- The insect pest can be controlled by application of any one of the insecticides given in Table 108.

Table 108. Insecticides used for controlling the Onion leaf–eating caterpillar

Sr. No.	Names of insecticides	Amount (per litre of water)
1.	Thiomethoxam 25 WG	1.0 g
2.	Acephate 75 SP	0.75 g
3.	Novaluron 10 EC	1.5 ml
4.	Cartap hydrochloride 50 SP	1.0 g

Diseases

1. Damping-off

C.O.: *Pythium, Rhizoctonia or Fusarium*

The damping–off disease is more prevalent in Northern and Eastern parts of the country during kharif season, causing 60–75 per cent damage.

Symptoms

Two types of symptoms are observed as follows:

Pre-emergence damping–off - The fungus kills the radicle and plumule of seed before emergence from the soil.

Post-emergence damping–off - The pathogen attacks the collar region of seedlings on the surface of the soil. Seedlings may fall over and die as a result of breakdown of plant tissues at the soil level.

Control measures

- The seeds that are free from disease should be collected from the field and should be treated with any one of the fungicides mentioned in Table 109 before sowing.
- Soil solarisation using polythene of 250 gauge for 10 d before sowing and application of Trichoderma viride in soil @ 1,250 g/ha is also effective to control damping off to a considerable extent.
- The disease can be controlled by spraying any one of the fungicides given in Table 109.

Table 109. Fungicides used for controlling the Damping off of Onion

Sr. No.	Names of fungicides/bio-fungicides	Amount (per litre of water)
1.	Carbendazim 50 WP	2.0–3.0 g
2.	M.E.M.C 6 WP	1.0 g
3.	Captan 50 WP	2.0–5.0 g
4.	Mancozeb 75 WP	2.5 g
5.	Trichoderma viridi	4.0–6.0 g

2. Onion Smut

C.O.: *Urocystis cepulae*

This disease produces fungus, which survives in soil where the temperature remains below 30°C.

Symptoms

Smut appears as elongated, dark, slightly thickened areas at the base of the seedlings. The black lesions appear near the base of scales on planting. The affected leaves bend downwards in an abnormal manner. Numerous raised blisters occur near the base of the leaves of older plants.

Control measures

- The disease can be controlled by treating the seeds with seed treatment chemicals.
- Seed–bed soil treatment with methyl bromide at 1 kg/25 m^2 or alternative application of formaldehyde solution @ 8 ml per litre of water per 10m^2 is effective in controlling the disease as a resistance management strategy.
- The proper crop rotation minimizes the infection of smut disease.

3. Botrytis leaf blight

C.O.: *Botrytis squamosa*

Botrytis leaf blight is a very common disease of Onion.

Symptoms

Oval–shaped spots of greyish–white colour having about 1–3 mm length appear on the leaves. These spots are often surrounded by a distinctive silvery–white 'halo' with uneven margins. The centres of many spots become sunken and straw–coloured and when numerous, the leaf tips begin to dieback, eventually affecting the entire leaf.

Control measures

- Cultural practices that promote air movement and rapid drying of leaves can help reduce infections and disease severity.
- Rotation with non-susceptible crops such as alfalfa or small grains suppresses the buildup of sclerotia in soil. Preventative application of fungicides when environmental conditions favour disease development can be an effective disease management tool.
- The disease can be controlled by spraying any one of the fungicides given in Table 110.

Table 110. Fungicides used for controlling Botrytis leaf blight

Sr. No.	Names of fungicides	Amount (per litre of water)
1.	Chlorothalonil 75 WP	2.0 g
2.	Mancozeb 75 WP	2.5 g
3.	Benomyl 50 WP	0.5 g

- Fungicides with activity against *Botrytis* include benomyl, dicloran, iprodione, mancozeb and maneb. Resistance to some fungicides has been reported, and any of these chemicals should be alternated with others.

4. Downy mildew

C.O.: *Peronospora destructor*

The disease often starts in patches and is favoured by cool (less than 22°C), humid weather. Downy mildew of onions is caused by the air–borne fungus.

Fig 33. Downy mildew disease in Onion

Symptoms

A purple–grey coloured, velvety growth first appears on otherwise green leaves (most easily seen in the early morning). Diseased leaves quickly turn pale–green, then yellow in colour, and collapse and die.

Control measures

It is possible to control this disease by spraying the crop with any one of the fungicides given in Table 111.

Table 111. Fungicides used for controlling the Downy mildew of Onion

Sr. No.	Names of fungicides	Amount (per litre of water)
1.	Mancozeb 75 WP	2.5 g
2.	Chlorothalonil 75 WP	2.0g
3.	Dithiolone 75WP	2.5 g
4.	Propineb 70 WP	3.0 g

Two or more of these materials should be rotated throughout the season as a resistance management strategy.

5. Onion powdery mildew (OPM)

C.O.: *Leveillula taurica*

Symptoms

The earliest infection symptoms appear as whitish lesion on older leaves and later when conditions are conducive, the fungus produces visible mycelium with prolific number of spores.

Fig 34. Powdery mildew in Onion

Control measures

The disease can be controlled by spraying any one of the fungicides given in Table 112.

Table 112. Fungicides used for controlling the Onion powdery mildew

Sr. No.	Name of fungicides	Amount (per litre of water)
1.	Sulphur 80 WP	2.0–3.0 g
2.	Azoxistrabin 50 WDG	1.0 ml
3.	Tridemorph 80 EC	1.0 ml
4.	Dithiolone 75 WP	2.5 g

5. Purple blotch

C.O.: *Alternaria porri*

Purple blotch, often appears on leaves that have already been damaged by other diseases or environmental factors. Warm (18–30°C) wet periods favour the development of purple blotch.

Symptoms

Small brown spots with purplish centres are characteristic of this disease. Under favourable conditions, the spots form into oval lesions that have a purplish tint with concentric rings. Older leaves tend to be more susceptible to the disease. Purple blotch causes Onion leaves to become blighted and die prematurely. The result is reduced yields from undersized and immature bulbs. In storage, diseased bulbs rot.

Control measures

- The seeds or bulbs should be collected from disease–free field.
- Overcrowding and application of excessive nitrogen should be avoided.
- The Onion should be cultivated in rotation.
- The disease can be controlled by spraying any of the fungicides given in Table 113 at 7–10 days interval.

Table 113. Fungicides used for controlling Purple blotch of Onion

Sr. No.	Names of fungicides	Amount (per litre of water)
1.	Copper oxychloride 50 WP	5.0 g
2.	Mancozeb 75 WP	2.5 g
3.	Carbendazim 50 WP	1.0 g
4.	Zineb 75 WP	2.5 g
5.	Chlorothalonil 75 WP	2.0 g

6. Basal Rot

C.O.: *Fusarium oxysporum* f. sp *Cepae*

Symptoms

The primary symptoms are yellowing, curling and necrosis at the tip of leaf blades. With time, whole leaf blades show symptoms and eventually wither and decay. Infected roots are dark brown in colour, flattened, transparent, and sometimes hollow.

Control measures

- Long–term rotation with non–host crops for four years or longer may also help to reduce losses.
- Dipping seedlings in any one of the fungicides solutions given in Table 114 before transplanting can also reduce the disease severity:

Table 114. Fungicides used for controlling the Basal rot of Onion

Sr. No.	Name of fungicide	Amount (per litre of water)
1.	Tricoderma viridi	4.0 g
2.	Carbendazim 50 WP	1.0 g
3.	Captan 50 WP	2.5 g
4.	Chlorothalonil 75 WP	2.0 g

7. Bacterial brown rot

C.O.: *Pseudomonas aeruginosa*

The bacterial brown rot is a very serious disease of Onions in storage.

Symptoms

The infection occurs through the wounds. The rot begins at the neck of the bulbs, which later gives foul smell through the neck when squeezed.

Control measures

- Proper curing and rapid drying of the bulbs after harvesting is essential for controlling the disease.
- Affected bulbs should be discarded before storage.
- If rains occur during maturity, it is recommended to spray Streptocycline (0.02%) on the plant.

8. Black mold

C.O.: *Aspergillus niger*

The black mold of Onion can occur in the field or during storage. Aspergillus in Onions is also common in extremely hot desert–like places where Onions are cultivated. Black mold is generally a post–harvest disease, although it may be seen on mature Onions in the field. The disease can be recognized by the presence of black powdery spore masses of fungus on the outer scales.

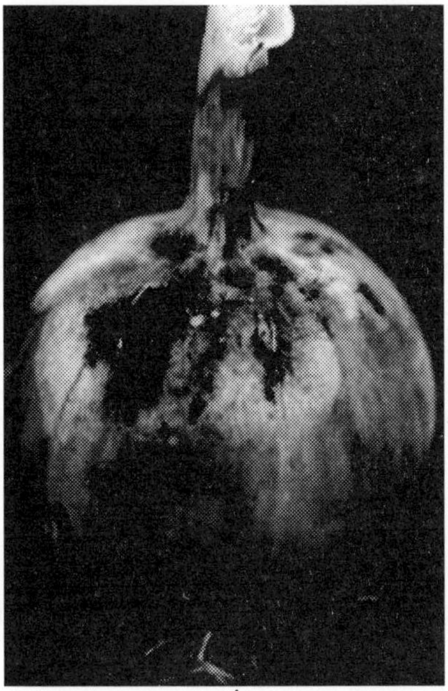

Fig 35. Black mold in Onion

Symptoms

As the common name suggests, the primary symptom is a black discolouration of tissue. Infected bulbs may show blackening at the neck, streaks or spots of black on or beneath the outer scales, and black discolouration in bruised areas. In advanced stages, the entire bulb appears black and shrivelled.

Control measures

- Bulbs should be cured thoroughly and rapidly before storage.
- Good ventilation should be maintained in the store house.
- Maintaining right storage temperature, just below 0°C, is an important measure in controlling this disease.
- Control of any foliar disease helps to reduce black mold during storage.
- It is also important to avoid wounding when bulbs are harvested, transported, or stored. Finally, proper drying and storage helps to prevent the occurrence of black mold.

9. Soft rot disease

a. Bacterial soft rot

C.O.: *Erwinia carotovora* subsp. *carotovora* and other species

Onions with mechanical injuries, Sunscald or bruises are particularly susceptible to bacterial soft rot, especially if they have been held under warm and humid conditions.

Symptoms

Soft rot is one of the more prevalent causes of loss in storage Onions. The soft rot bacterium can enter the neck tissues as plants approach maturity. In the field, plants wilt and die. As the rot progresses, invaded scales become soft and foul-smelling. Infected bulbs have a pungent smell and they rot very rapidly.

b. Fusarium rot

C.O.: *Fusarium* sp

Generally, Fusarium rot causes minor losses, but rotten Onions in a bag can deteriorate the quality of other Onions. In warm climates of 28°C–32°C, the rot normally starts in the field at the base of the Onion and progresses into the bulb causing a soft watery rot.

Control measures

- The damaged bulbs, either mechanically or by insects, should be discarded before storing.
- Thorough and rapid drying of bulbs at harvesting time is essential for controlling this disease.
- Good ventilation in the storage should be maintained.

10. Aster yellow

C.O.: *Aster yellow phytoplasma*

The pathogen is transmitted by the aster leafhopper (*Macrosteles quadrilineatus* Forbes).

Symptoms

The symptoms initially appear as yellowing at the base of young leaves, which then spreads towards the top. Leaves flatten, become marked with green and yellow streaks, but do not twist. Flower stems are abnormally elongated, and have malformed, sterile floral clusters.

Control measures

- Crop rotation and elimination of weed hosts in and around Onion fields help in reducing the disease incidence.
- The disease–infected parts of the plant should be destroyed.
- The insect vector should be controlled by spraying any one of the insecticides given in Table 115.

Table 115. Insecticides used to control the insect vector of Aster yellow of Onion

Sr. No.	Name of insecticide	Amount (per litre of water)
1.	Acephate 75 WP	0.75 g
2.	Quinalphos 25 EC	2.0 ml
3.	Clothianidin 50 WDG	0.1 g
4.	Triazophos 40 EC	2.0 ml

11. Nutrient deficiencies

Onions require a well–balanced nutrient regime and if certain nutrients are low, some symptoms are seen as follows:

Table 116. Nutrient deficiencies of Onion

Sr. No.	Name of deficient nutrient	Nutrient deficiency symptoms
1.	Nitrogen (N)	Pale leaves
2.	Phosphorus (P)	Pale leaves
3.	Manganese (Mn)	Yellowing of leaves between leaf veins
4.	Zinc (Zn)	Stunted, twisted and yellow–striped foliage

Control measures

- **Nitrogen** - Application of nitrogenous fertilizers, particularly urea as foliar spray @ 20 g/l of water
- **Phosphorus** - Application of phosphorus fertilizer
- **Manganese** - Application of Manganese sulphate @ 16.25–33.25 kg/ha or 6 g/l of water
- **Zinc** - Application of Zinc sulphate @ 2.0–22.4 kg/ha, according to soil testing report or 6 g/l of water.

15 CELERY

Scientific name - *Apium graveolens*

Family - Apiaceae

Commercial part - Seed

Indian names - *Bengali* : Bandhuri, Chanur; *Hindi* : Shalari, Ajmud, Randhuni; *Gujarati* : Bodiajmoda; *Malayalam* : Ayamodakam; *Marathi* : Ajmoda; *Punjabi* : Kernauli; *Sanskrit* : Ajmoda; *Tamil* : Ajmoda

Apium graveolens is a plant species in the family Apiaceae, commonly known as Celery (var. *dulce*) or celeriac (var. *rapaceum*), depending on whether the petioles (stalks) or roots are eaten.

The native habitat of Celery extends from Sweden to Egypt, Algeria and Ethiopia, and in Asia - India and China.

Fig 36. Celery

15.1. BOTANICAL DESCRIPTION

- **Plant** - It is an herbaceous plant, usually 60–120 cm in height. It is an annual plant in the planes and biennial in cold climate and on the hills.
- **Leaf** - The leaves are oblong, 7–18 cm in size, long, pinnate or trifoliate.
- **Root** - The roots are succulent, well developed with numerous lateral roots.
- **Flower** - The flowers are white in colour and borne in compound umbels.
- **Stems** - The stems are branched, angular, jointed and light green in colour.
- **Fruit** - The fruit is made up of two united umbels, each containing a seed.
- **Seed** - The seeds are very small in size, about 1-2 mm in length, oval and greenish brown in colour. The Celery seed has a pleasingly crisp texture and subtle flavour, but a slightly bitter taste.

15.2. ORIGIN AND DISTRIBUTION

The place of origin of Celery extends from Sweden, southward to Algeria, Egypt, Abyssinia, and in Asia even to the Caucasus, Baluchism and the mountains of India. In India quite a large area in the Punjab and Uttar Pradesh is diverted to the production of the seeds. These seeds are being exported to European countries as a condiment. It is cultivated at the foothills of north western Himalayas and the outlying hills of Punjab, Himachal Pradesh and Uttar Pradesh. It is mostly produced in Punjab (Amritsar, Gurdaspur and Jalandhar) followed by Uttar Pradesh (Ladhwa, Saharanpur districts).

15.3. CLIMATE

Celery is a Rabi season crop. It requires warm days and cool nights, low humidity and plenty of sunshine. It is a moisture–loving plant that requires long, cool seasons with well–distributed moderate rainfall during the growing season. It requires cool climate during growth stage and warm dry climate during maturity. A week long cold spell during mid seasons can induce bolting and lower seed yield.

15.4. SOIL

Celery can be cultivated in almost all types of soils, but the crop thrives well on silt loam to loamy soil (pH 6–8.5) with moderate organic matter. Clayey soils and water logging conditions are not suitable for this plant. Rich, friable, well–drained and deep soil is good for the cultivation of Celery.

15.5. SOWING OF SEED

1. Time of sowing : Mid–September to Mid–November

2. Seed rate: 2.0 kg/ha

3. Method of sowing seeds

The seeds are sown in nursery by broadcasting method or in lines by hand in the small furrows made by hand or by a stick or khurpi. The nursery should be raised in beds of 3 x 1 m size in about 200 sq m area for transplanting one hectare land. Manuring schedule for seed bed (200 sq m area) is as follows:

Manures

Compost or FYM: 500.0–600.0 kg

Fertilizers

Nitrogen (N)	: 1.6 kg
Phosphorus (P_2O_5)	: 0.8 kg
Potassium (K_2O)	: 0.4 kg

15.6. PLANTING

The seedlings (12–15 cm tall and 60 d old) with 7–8 cm root growth should be transplanted in the field to ensure a good crop stand. This enables the crop to withstand early season frost and low temperature. Flowering takes place during the month of March and seeds mature by the end of April. It remains in the field for a period of 120–140 d.

Spacing: 30 x 30 cm or 40 x 25 cm

15.7. MANURING

Manures

Compost or FYM: 25.0–30.0 t/ha

Fertilizers

Nitrogen (N)	: 80.0 kg/ha
Phosphorus (P_2O_5)	: 40.0 kg/ha
Potassium(K_2O)	: 20.0 kg/ha

In some tracts calcium, magnesium and boron are needed and they should be applied after testing the soil.

Method of application

Manures

The manures should be applied at the time of land preparation and mixed well in the soil by ploughing followed by planking.

Fertilizers

Basal dressing - Total amount of phosphorus and half amount of nitrogen to be applied as basal dressing.

Top dressing - The remaining amount of nitrogen should be applied 30 DAT.

15.8. INTERCULTURAL OPERATION

The land should be kept free from weeds. To keep the weeds under check, hoeing should be done for 3–4 times. The first intercultural operation or hoeing should be done 20–25 days after transplanting. Subsequently, intercultural operations should be repeated thrice based on the need at an interval of 20–25 days.

15.9. IRRIGATION

Celery is a shallow–rooted crop and its moisture requirements are high. The first irrigation should be applied on the day of transplanting and subsequent irrigations should be given at fortnightly intervals and weekly nearing seed maturity.

15.10. HARVESTING

Harvesting of plants should be done when about 80% umbels turn to light brown colour. The harvested plants should be dried on clean threshing yard and threshing should be done by beating with sticks. Winnowing should be done by traditional method to clean seeds. The seeds should then be dried in bright sunlight. The seeds should be kept in gunny bags or stored in polythene bags.

15.11. YIELD

1,000–1,200 kg/ha

15.12. VARIETIES

Giant Pascal, Summer Pascal, Emperor of Jeen, Golden Self Blanching, Fordhook, Standard bearer, Wright Glove Grant, EC-99249-1, PRL-85-1, Wonderful or Golden Plume, Michigan, Detroit Golden, Flora-55, Flora-88, Red Celery etc.

15.13. PLANT PROTECTION

No major insect pests and diseases attack Celery.

16 NUTMEG

Scientific name - *Myristica fragrans*

Family - Myristicaceae

Commercial part - Seed (Kernel) and mace

Indian names - *Bengali* : Jaiphal; *Hindi* : Jaiphal; *Gujarati* : Jaiphal; *Kannada* : Jayikai; *Kashmiri* : Zaaphal; *Malayalam* : Jathikka; *Marathi* : Jaiphal; *Oriya* : Jaiphala; *Punjabi* : Jaiphal; *Sanskrit* : Jatiphala; *Tamil* : Jathikai; *Telugu* : Jajikai; *Urdu* : Jaiphal.

Nutmeg, a large evergreen tree, is indigenous to Moluccas Islands and now is cultivated in West Indies. Nutmeg produces two spices namely Nutmeg and mace. Nutmeg is the seed kernel inside the fruit and mace is the dried aril surrounding it.

In India, Nutmeg is mainly cultivated in Thrissur, Ernakulam and Kottayam districts of Kerala and parts of Kanyakumari and Tirunelveli districts in Tamil Nadu.

16.1. CLIMATE

Nutmeg requires a hot, humid climate without pronounced dry season and it thrives well in locations with an annual rainfall of 150 cm and more. The tree prefers partial shade. It grows well from about 1,300 m above MSL.

16.2. SOIL

The Nutmeg grows in any type of soil. But the soil rich in organic matter and well drained is best suited for the cultivation of Nutmeg. Clay loam, sandy loam, red and laterite soil are ideal soils for Nutmeg cultivation.

1. Seed

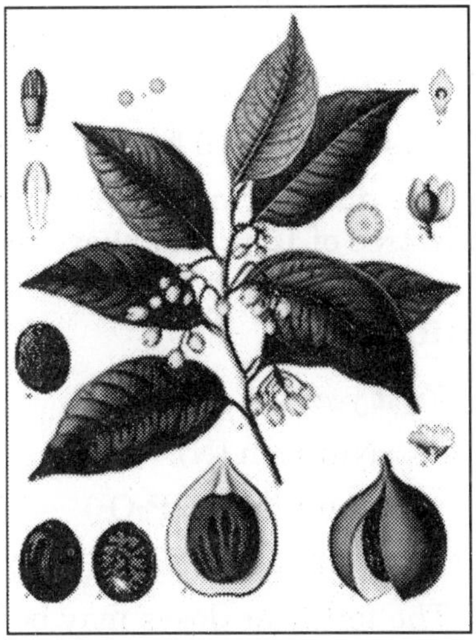

2. Plant

Fig. 37. Nutmeg - 1. Seed 2. Plant

16.3. LAND PREPARATION

The soil should be brought to fine tilth by ploughing the soil with tractor and power tiller. The manure should be applied at the time of land preparation and mixed with soil by ploughing followed by planking.

16.4. PROPAGATION

Nutmeg is propagated by planting seedlings. The seedlings should be raised in a nursery or in polythene bags. Fully ripened tree–burst fruits should be selected for raising seedlings. The fleshy rind and the mace should be removed before sowing. The seeds should be sown immediately after collection in seed beds of 100–120 cm width, 15 cm height. These seed beds should be of convenient length, and may be prepared in cool and shady places. A mixture of garden soil and sand in the ratio 3:1 may be used for preparing nursery beds. The seeds should be sown by dibbling method at a spacing of about 12 cm on either side.

16.5. PLANTING

Pits of 60 x 60 x 60 cm should be dug in the well–prepared land at a spacing of 8 x 8 m. Pits should be refilled with mixture of soil and compost in the ratio 3:1. Seedlings should be planted carefully in the centre of the pits during rainy seasons. Young plants should be provided with artificial shade and irrigated during summer months.

16.6. MANURING

Manures

First year - 10.0 kg per seedling

The quantity of manures may be increased gradually every year till a well-grown tree of 15 yr and above receives 50 kg of organic manures per year.

Fertilizers

First year -

Nitrogen (N)	: 20 g/plant
Phosphorus (P_2O_5)	: 18 g/plant
Potassium (K_2O)	: 50 g/plant

The fertilizer doses may be increased gradually every year till a well–grown tree of 15 yr or more receives about 500 g nitrogen (N), 250 g phosphorus (P_2O_5) and 100 g potash (K_2O) annually. The fertilizers should be applied in two equal split doses, first in May–June along with organic manure and the other in September–October.

16.7. IRRIGATION

Irrigation is necessary soon after planting seedlings in the main field and also during summer seasons. At other times, irrigation is done as and when required.

16.8. HARVESTING

The fruits are either plucked from the tree or allowed to drop. The nuts are dried in the Sun for six to eight days till they rattle in their shell. They are stored in warm dry place prior to shelling. Dried Nutmeg and mace are directly used as spice and also for the preparation of their derivatives.

16.9. YIELD

Nutmeg - 800 kg /ha

16.10. VARIETIES

IISR-Viswashree etc.

16.11. PLANT PROTECTION

Insect pests

No serious insect pests attack the Nutmeg.

Diseases

1. Die–back

C.O.: *Diplodianatalensis* sp

The die–back is a serious disease in Nutmegs.

Symptoms

The mature and immature branches of Nutmeg dry from tips to downwards.

Control measures

- The infected branches should be removed and destroyed by burning.
- The Bordeaux paste or paste made from Copper oxychloride 50 WP should be smeared on the cut end of the branches.

2. Fruit rot

C.O.: *Diplodia natalensis* and *Phytophthora* sp

The fruit rot is an important disease in Nutmegs.

Symptoms

The disease–affected fruits get split and subsequently lead to rotting of mace.

Control measures

- The rotted fruits should be collected from the plants or fallen fruits and destroyed by burning.
- The disease is possible to be kept under control by spraying the plants with any one of the fungicides given in Table 117.

Table 117. Fungicides used for controlling Fruit rot of Nutmeg

Sr. No.	Name of fungicide	Amount (per litre of water)
1.	Mancozeb 75 WP	2.5 g
2.	Carbendazim 50 WP	1.0 g
3.	Copper oxychloride 50 WP	5.0 g

3. Leaf spot and shot hole

C.O.: *Colletotrichum gloeosporioide*

Symptoms

Sunken spots surrounded by a yellow halo are the initial symptoms of this disease. Subsequently, the central portion of the necrotic region drops off resulting in shot hole symptoms.

Control measures

The disease can be controlled by spraying the plant with any one of the fungicides given in Table 118.

Table 118. Fungicides used for controlling Leaf spot and shot hole of Nutmeg

Sr. No.	Name of fungicide	Amount (per litre of water)
1.	Mancozeb 75 WP	2.5 g
2.	Carbendazim 50 WP	1.0 g
3.	Copper oxychloride 50 WP	5.0 g
4.	Thiophenate methyl 70 WP	1.0 g
5.	Metalaxyl 35 WS	0.5 g

17 ALLSPICE

Scientific name - *Pimenta dioica* (L.) Merr.

Family - Myrtaceae

Commercial part - Unripe fruit (Berry)

Common name(s) - Allspice, pimenta, Jamaica pepper, clove pepper, pimento

Allspice, also called Jamaica pepper, pepper, myrtle pepper, pimenta or newspice, is the dried unripe fruit (berries) of *Pimenta dioica*, a mid-canopy tree. The fruits are picked when green and unripe, and traditionally, dried in the Sun. Allspice is used in pickles, ketchup and marinades, and to flavour pumpkin pies, cakes and candies. Oil extracted from the fruits is used in perfumes and cosmetics.

Fig 38. Allspice fruit

17.1. ORIGIN AND DISTRIBUTION

Allspice is native to the Greater Antilles, southern Mexico and Central America, now cultivated in many warm parts of the world. The name 'Allspice' was coined as early as in 1621 by the English, who thought it combined the flavours of Cinnamon, Nutmeg and Cloves.

It was discovered in Mexico by 16th century Spanish explorers who called it 'pimienta', confusing it with black pepper. Nowadays Allspice is grown commercially in Mexico, Honduras, Trinidad, Cuba and especially in Jamaica, which practically has a monopoly.

17.2. BOTANICAL DESCRIPTION

- **Tree** - The tree is small in size and evergreen, and grows to 7–13 m height with large leaves measuring 4–8 inches (10–20 cm).
- **Leaf** - The tree has leathery, smooth, grey bark, and has elliptic, glossy leaves that are dark green in colour. The leaves are up to 15 cm long, opposite, oblong, aromatic and quite attractive.
- **Flower** - The flowers are white in colour and are about 0.25 inches (0.6 cm) across. They are borne in many flowered pyramidal cymes originating from the leaf axils. Flowers are structurally hermaphrodite, but functionally dioeciously.
- **Fruit** - The fruit is a brown, berry–like drupe, about 0.25 inches (0.6 cm) long. Fruit is a two–seeded berry. Male trees flower earlier. The leaves and fruits smell like a combination of Cloves, Black pepper, Nutmeg and Cinnamon. Hence, the common name is Allspice.

17.3. CLIMATE

Allspice requires warm humid climate. It thrives in all situations ranging from sea level up to an altitude of 1,000 m. Allspice grows in an area where it receives sunlight for at least 40 per cent of the day. Younger Allspice trees need more light to remain healthy.

17.4. SOIL

Deep loam soil with high humus content found in the forest region is best suited for its cultivation. It grows satisfactorily on laterite soil, loam and rich black soil having good drainage.

17.5. PROPAGATION

Allspice is propagated through seeds. The seeds should be sown in raised beds of 15–20 cm height, 1 m width and convenient length. The beds may be made of loose soil sand mixture having sand in the top layer. The seeds germinate in 15 d and continue up to 40–45 d. The seedlings can be transferred to bags three weeks after their emergence above ground level. The seeds can be stored as ripe berries after collection without extracting seeds up to three weeks. But viability gets reduced slowly after a period of three weeks and is lost completely after nine weeks.

17.6. PLANTING

1. Time of planting : June–July

2. Method of planting

The slender and delicate plants are planted in a pit measuring 75 x 75 x 75 cm. The pits are partially filled with compost, green leaf or cattle manure and covered with top soil. The grown seedlings (six months old) are planted in the field with spacing as follows:

Row to row : 6 m

Plant to plant : 6 m

17.7. MANURING

The judicious manuring is found good for getting higher yield. The amount of manures varies with the age of the plant as follows:

Table 119. Recommended amount of manures and plant nutrients for Nutmeg

Sr. No.	Age of the Plant (y)	Manures (kg/plant)	Amount of nutrient (g/plant)		
			Nitrogen (N)	Phosphorus (P_2O_5)	Potassium (K_2O)
1.	1	15.0	20	20	60
2.	2	30.0	40	35	100
3.	15	50.0	300	250	750

In addition to this, 50 g each of Azospirillum and Phosphobacterium should be applied one month after manuring.

- **Method of application**

Manures

Manures should be applied in the month of May–June, with the commencement of monsoon.

Fertilizers

Fertilizers should be applied in two equal split doses i.e., one during the months of May–June and other during the months of September–October.

Manures and fertilizers need to be applied in the trench that is dug around the tree about 50–160 cm away from the base depending upon the age.

17.8. INTERCULTURAL OPERATION

1. Weeding

The field of Allspice should be kept free from weeds during early stages of crop growth. For this, regular weeding should be done at regular intervals. Dead and diseased shoots should be removed once or twice a year.

2. Pruning

The pruning of Allspice tree should be done during the spring to summer months. An Allspice tree grown outside in ideal conditions reaches around 12 m height.

17.9. HARVESTING

Plants start flowering when they are 7–10 years old, but the peak harvest is obtained from 15–20 years. The berries should be harvested when fully grown, but still green. They should be sun–dried for about 3–12 days and should be frequently turned down with ladles, till the berries are completely dried. Curing is complete when the berries become crisp and produce a metallic sound when shaken.

17.10. YIELD

Dry berries - 50–60 kg/ plant.

18 CURRY LEAF

Scientific name - *Murraya koenigi*

Family - *Rutaceae*

Commercial part - Leaf

Indian names - *Bengali* : Barsanga, Kariphulli, Karipatta; *Assamese* : Narsinghs, Bisharhari; *Hindi* : Kathnim, Mitha neem, Curry or kurry patta, Gandhela, Bareanga; *Gujarati* : Goranimb, Kadhilimbdo; *Kannada* : Karibevu; *Malayalam* : Karriveppilei; *Marathi* : Karhinimb, Poospala, Gandla, Jhirang; *Oriya* : Barsan, Basango, Bhuraunga; *Punjabi* : Curry patta; *Sanskrit* : Krishna nimbi; *Tamil* : Karivempu, Karuveppilei; *Telugu* : Karepaku.

Curry leaf is an important perennial tree. Its leaves are used mainly to improve the taste and flavour of the food.

18.1. ORIGIN AND DISTRIBUTION

Curry leaf is native to India. Curry leaf is found almost throughout India up to an altitude of 1,500 m. It is much cultivated for its aromatic leaves.

18.2. BOTANICAL DESCRIPTION

- **Plant** - It is a small tree, growing 4–6 m (13–20 ft) tall, with a trunk up to 40 cm diameter.
- **Leaf** - The leaves are pinnate with 11–21 leaflets, each leaflet 2–4 cm long and 1–2 cm broad. They are highly aromatic.
- **Flower** - The flowers are small, white and fragrant. The small black shiny berries are edible, but their seeds are poisonous.

Fig. 39. Curry leaf twing

18.3. CLIMATE

Curry leaf can tolerate temperature up to 37°C, but its growth is affected below 16°C. It requires about 20°C to germinate and may take a long time to germinate. Established plants need a sunny, well–drained place. The optimum temperature requirement is 26°C– 37°C.

18.4. SOIL

Curry leaf can be cultivated in a wide range of soils. Red sandy loam soil is ideal for its cultivation.

18.5. PROPAGATION

Curry leaf is mainly propagated through seeds. For raising seedlings, well–ripe fruits should be collected from high–yielding plants. The seeds should be sown either in nursery or polythene bags filled with a mixture of 1:1:1 sand, soil and FYM. Seeds germinate in three weeks.

18.6. PLANTING

The soil should be brought to a fine tilth by ploughing with tractor and powertiller followed by planking. Planting in the main field should be done during the monsoon season. Pit of size 30 x 30 x 30 cm should be dug one to two months before planting at a spacing of 1.2–1.5 m.

One year old seedlings are suitable for planting. One seedling should be planted at the centre of the pit in the main field. It can also be propagated by root suckers. There are a number of root suckers near its plants. They should be separated from the main plant during rainy season and planted immediately in the main field. The main field should be ploughed thoroughly. FYM @ 10 kg/pit should be applied.

18.7. MANURING

The amount of manures and fertilizers that are to be applied per plant per year are as follows:

Manures: Compost or FYM: 10.0 kg

Fertilizers

Nitrogen (N)	: 60 g
Phosphorus (P_2O_5)	: 80 g
Potassium (K_2O)	: 40 g

The fertilizers may be applied at the onset of the monsoon season.

18.8. INTERCULTURAL OPERATION

1. Weeding

The field of Curry leaf should be kept free from weeds. For this, weeding should be done at early growth stage of the crop.

2. Training and pruning

The plants should be trained and pruned to maintain a height of 1 m. The terminal buds of Curry leaf should be removed to encourage lateral branching. A minimum of 5–6 branches should be kept per plant.

18.9. IRRIGATION

The crop should be irrigated immediately after planting. Second irrigation should be given on the third day, and then the irrigation should be given once in a week.

18.10. HAVESTING

Curry leaves should be picked 15 months after planting. Commercial harvest can be started with three year old plants. With good management, normal yield can be obtained up to the age of 20–25 yr. The leaves can be harvested at 2½ –3 months intervals.

18.11. YIELD

Green leaves - 20.0–25.0 ton/ha

18.12. VARIETIES

Sen Kaampa, Dharwad-1, Dharwad-2 etc.

18.13. USES

In the kitchen, the curry leaves are used for a warm, appetizing aroma and a subtle, spicy flavour with meat, seafood or vegetable curries, chutneys, pickles, coconut sauces, relishes, omelettes, marinades, and vegetarian cuisine.

19 GARCINIA

Scientific name - *Garcinia cambogia*

Family - Cluciaceae

Commercial part - Fruit

Common names - Malabar tamarind, Brindall berry, Chikana red mango, Mangosteen, Vrikshamla, Kankushta

Garcinia is native to Asia, Australia, tropical and southern Africa and Polynesia. *Garcinia cambogia*, an exotic fruit grown in South India, has been used to impart a distinctive sour flavour to Indian cooking. The dried thick outer coat of its fruit is rich in acids and has been valued for its marked antiseptic properties. The fruit of *Garcinia cambogia* has been traditionally used in food preparation and cooking.

Fig. 40. Garcinia fruit

It is found in Maharashtra, Goa, Karnataka, Kerala, South Gujarat, Assam and West Bengal.

19.1. CLIMATE

Garcinia can be grown in different agro–climatic conditions.

19.2. SOIL

Garcinia can be grown in a variety of soils, but sandy loam, loam and clay loam soils are best suited for its cultivation.

19.3. PROPAGATION

Garcinia can be propagated by seed and by grafting. It can be cultivated by soft wood grafting and planted during the months of July–August. For raising seedlings, well–ripe fruits should be collected from high–yielding plants. The seeds should be sown either in nursery or polythene bags filled with a mixture of 1:1:1 sand, soil and FYM. The crop can be raised as a pure or as a perennial intercrop of coconut, areca nut etc. Pits of the size 75 x 75 x 75 cm in hard and laterite soils, 50 x 50 x 50 cm in sandy and alluvial soils should be prepared. The spacing should be 4 x 4 m for grafting and 7 x 7 m for seedlings. Before planting, pits should be filled with top soil and 2 kg of compost or well–decomposed cattle manure and 10 g of Carbaryl dust to avoid white ants attack. Generally, planting should be done during the months of July–October.

19.4. INTERCULTURAL OPERATION

1. Weeding

The field should be kept free from weeds. Weeding may be done as and when required usually 15–20 DAP.

2. Pruning

To keep the balance of the weight of the plant, the fast–growing branches of the plant should be pruned from third year onwards.

19.5. IRRIGATION

The plant should be irrigated as per need of the plant. Water stagnation in the field should be avoided.

19.6. MANURING

Manures

Compost or FYM: 10.0 kg per seedling or graft during the first year.

The quantity should be gradually increased till a well–grown tree of 15 yr and above receives 50 kg of organic manure per year.

Fertilizers

The amount of different plant nutrients to be applied per plant during the first year is as follows:

Nitrogen (N)	: 20 g
Phosphorus (P_2O_5)	: 18 g
Potassium (K_2O)	: 50 g

The above–mentioned quantity should be doubled in the second year and gradually increased to 500 g N, 250 g P_2O_5 and 1,000 g K_2O per plant per year so as to obtain full dose from 15th year onwards.

19.7. YIELD

Ripe fruit - 8.5 ton/ha

19.8. USES

The small pumpkin–shaped fruit has been used for centuries by many cultures worldwide for treatment of weight loss, but has recently become readily available to those in the West also.

The dried rind of *Garcinia cambogia*, commonly known as Malabar tamarind, has unique and varied use as condiment for flavouring curries, curing and drying fish, and used as medicines for various diseases.

20 MUSTARD AND RAPESEED

Scientific name - *Brassica juncea* (L.) Czern. & Coss

Family - Brassicaceae

Commercial part - Seed and leaf

Indian names - *Bengali* : Sarisha; *Hindi* : Rai, Banarasi rai, Kalee sarson; *Gujarati* : Rai; *Marathi* : Moharee, Rai; *Kannada* : Sasave; *Kashmiri* : Aasur, Sorisa; *Malayalam* : Kaduku; *Punjabi* : Rai, Banarasi rai, Kalee sarson; *Sanskrit* : Asuri, Bimbata; *Tamil* : Kadugo; *Telugu* : Avalu; *Urdu*: Rai, Banarasi rai, Kalee sarson.

Mustard is a condiment made from the seeds of a Mustard plant. Mustard is an annual herb cultivated as oilseed crop or as vegetable or as fodder. In India, the Mustard and rapeseeds the most important oilseed crop after groundnut, accounting around 25 per cent of total oilseed production. It is one of the important oilseed crops of the Indo–Gangetic plains.

20.1. BOTANICAL DESCRIPTION

- **Plant** - The plant is an annual herb with long and tapering root.
- **Leaf** - The leaves are stalked, glabrous or hairy, and bright green in colour.
- **Inflorescence** - Corymbose raceme
- **Flower** - The flowers are regular, bisexual, hypogynies, and syncarpous. Flower is yellow to pale yellow in colour with four sepals, four petals and tetradynamous stamens.
- **Fruit** - Siliqua
- **Seed** - The seed is small, round and black or white in colour, based on the variety.

Table 120. Classification of Mustard - rapeseed

Sr. No.	English name	Vernacular name	Botanical name	Identification characteristics of seeds
1.	Indian mustard / Brown mustard	Rai, Ryada, Raya, Laha, Lahta, Sasve, Herbo	*Brassica juncea* (L.) Czern. & Coss.	Seeds are medium in size, round and dark brown or black in colour.
2.	Indian mustard / Rapeseed/Toria	Toria, Tori, Lahi	*Brassica rapa* L. var. *Toria* (Syn. *B.campestris* L. var. *Toria*)	Seeds are dark brown, bold and large in size.
3.	Brown sarson / Rapeseed	Brown sarson, Bhoori sarson	*Brassica rapa* L. var. Brown sarson (Syn. *B. campestris* L. var. Brown sarson)	Seeds are light reddish in colour, bold, large in size.
4.	Yellow sarson / Colza/Rapeseed	Yellow sarson, Pilli sarson	*Brassica rapa* L.var. Yellow sarson (Syn. *B. campestris* L. var. Yellow sarson)	Seeds are slightly smaller in size than sarson, ovoid in shape, and yellow in colour.
5.	Rapeseed	Gobhi sarson	*Brassica napus* L.	Seeds are brownish black and large in size.
6.	Abyssinian Mustard / Ethiopian mustard	Karan rai	*Brassica carinata* A. Br.	Seeds are small, round and reddish brown in colour.
7.	Rocket salad	Duan, Tera, Tara, Saundh, Taramira	*Eruca sativa* Mill	Seeds are light reddish brown in colour and distinctly ovoid in shape.

Source: 1. National Research Centre on Rapeseed-Mustard, Sewer, Bharatpur (Raj.);
2. Department of Botany, Nagpur University

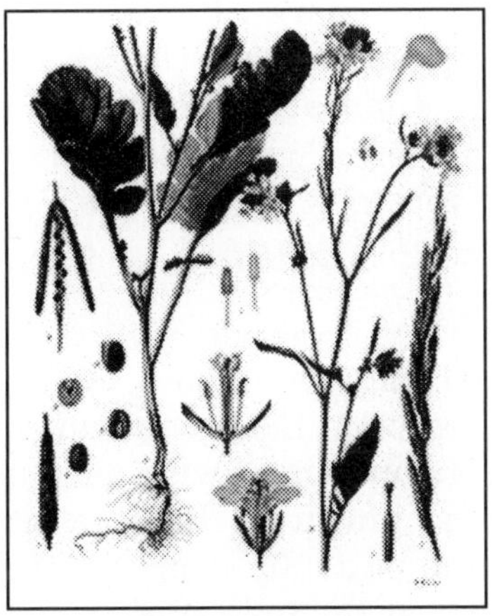

Fig 41. Mustard (*Brassica nigra*)

Fig 42. Brown or Indian mustard (*Brassica juncea*)

Fig 43. Sarson (*Brassica campestris*)

20.2. ORIGIN AND DISTRIBUTION

The yellow and white Mustard is indigenous to southern Europe, whereas Brown mustard is from China, and introduced to northern India. The Black mustard is endemic in the southern Mediterranean region. The White mustard is widely cultivated in Australia, China, Chili, Denmark, Italy, Japan, UK, Netherlands, North Africa, Canada, and USA. Indian mustard (*Brassica juncea*) is cultivated in the States of Assam, Bihar, Gujarat, Haryana, Himachal Pradesh, Jammu & Kashmir, Madhya Pradesh, Orissa, Punjab, Rajasthan, Uttar Pradesh and West Bengal. It is cultivated as a Rabi crop.

20.3. CLIMATE

Mustard is the crop of tropical as well as temperate zones and requires somewhat cool and dry weather for satisfactory growth. Cool temperature, clear dry weather with plentiful of bright sunshine accompanied with adequate soil moisture increases the oil yield. In India, it is grown in Rabi season from September–October to February–March. It requires a fair supply of moisture during the growing period and a dry clear weather at the time of maturity. These crops are not drought tolerant. They require an annual precipitation of 35–45 cm and moderate temperature of 21°C–25.7°C for their growth and development.

20.4. SOIL

Mustard is capable of growing under a wide range of soil conditions varying from sandy loam to clay loam soils, but thrives best on light loam soils. Rai may be grown on all types of soil, but does best in loam to heavy loam soil. Sarson is suited in light loam soil and Taramira is mostly grown on very light soil. The soil should have good drainage as these crops cannot withstand waterlogging condition. The crop grows well in a soil having pH of 6.0–7.5.

20.5. LAND PREPARATION

Mustard and rapeseed are small–seeded crops. Hence, a fine seed bed is required to ensure good germination. Mustard and rapeseed are grown as pure crop and mixed crop with gram (*Cicer arietinum*), wheat (*Triticum* sp), lentil (*Lens esculentus*), radish (*Raphanus sativus*) etc. and accordingly land should be prepared as follows:

1. Mixed crop

While growing Mustard–rapeseed as mixed crop, the field should be prepared with a view to meet the requirement of the main crop.

2. Pure crop

The land should be well prepared by 4–6 ploughings with tractor or power tiller followed by planking after each ploughing. The land should be made free from weeds and stubbles of previous crop. All the clods of the soil should be broken and the land should be made pulverized, loose and friable. The manures should be applied during land preparation and mixed well with the soil by ploughing followed by planking.

20.6. SOWING OF SEED

1. Time of sowing

The time of sowing plays an important role in the yield of this crop. The optimum time for sowing Mustard and rapeseed recommended for different States is given Table 121.

Table 121. Time of sowing for different States

S.N	State	Rai (Mustard)	Toria	Brown sarson	Yellow sarson
1.	Assam	-	Mid–October to Mid–November	-	-
2.	Bihar and Jharkhand	Second fortnight of October	Last week of September to first week of October	-	First fortnight of October

3.	Gujarat	Second fortnight of October	-	-	Second week of October
4.	Haryana	Mid–October	End of August to Mid–September	End of October to early November	End of September to early October
5.	Madhya Pradesh	Early October	Second fortnight of September	-	-
6.	Punjab	Second fortnight of October to first fortnight of November	First fortnight of September	End of September to early October	End of September to early October
7.	Rajasthan	Mid–September	End of August to beginning of September	Mid–September	Mid–September
8.	Uttar Pradesh and Uttaranchal	First fortnight of October	Early to mid–September	-	-
9	West Bengal	First week of October to second week of November	First week of October to last week of October	-	Mid–October to first week of November

2. Seed treatment

Seed should be healthy and free from diseases. The seed should be treated with any one of the fungicides given in Table 122 to protect the crop from seed–borne diseases:

Table 122. Fungicides used for seed treatment of Mustard and rapeseed

S.N.	Name of fungicide/bio-fungicide	Amount (per litre of water)
1.	Carbendazim 50 WP	2.0–3.0 g
2.	M.E.M.C 6 WP	1.0 g
3.	Captan 50 WP	2.0–5.0 g
4.	Mancozeb 75 WP	2.5 g
5.	Trichoderma viridi	4.0–5.0 g

3. Seed rate

The seed rate varies according to the method of sowing as follows:

Pure crop: 6.25–7.5 kg/ha

Mixed crop: 2.50–3.75 kg/ha

4. Method of sowing

Seed is generally sown by broadcasting method, but line sowing with the help of seed drill or behind the plough has been found to be very good and economical. The pure crop of Mustard and rapeseed is sown in lines, maintaining a spacing of 30 cm between rows and 8–10 cm between plants, and at a depth of 4–5 cm. Whenever moisture in the field is inadequate, the seed is mixed with moist soil and kept overnight or soaked overnight in water to get uniform germination. For even distribution, the seed is usually mixed with sand before sowing. In mixed cropping, the seeds are sown in rows, 1.8–2.4 m apart across the main field.

5. Spacing

Mustard	: 30–45 cm x 10–20 cm
Toria	: 30–40 cm x 10–15 cm
Brown sarson	: 30 x 10 cm
Yellow sarson	: 30 x 10 cm

20.7. MANURING

The rapeseed and Mustard crop exhaust heavy amount of nutrients from the soil. Soil testing must be done to assess the fertility status and the fertilizer doses must be decided accordingly. The recommended doses of manures and fertilizers (kg ha^{-1}) are as follows:

Manures

Compost or FYM: 5.0 t/ha

Azophos: 15.0 kg/ha

Fertilizers

The State–wise fertilizer recommendation is given in Table 123.

Table 123. State–wise fertilizer recommendations - Mustard and rapeseed

Sr. No.	State	Condition	Mustard			Toria			Brown/Yellow Sarson		
			N	P_2O_5	K_2O	N	P_2O_5	K_2O	N	P_2O_5	K_2O
1.	Assam	Rainfed	-	-	-	40	35	15	-	-	-
2.	Bihar and Jharkhand	Irrigated	80	40	40	50	25	25	60	40	40
		Rainfed	40	20	20	40	30	20	30	20	20
3.	Gujarat	Irrigated	50	50	-	-	-	-	60	50	50

4.	Haryana	Irrigated	60	30	-	60	-	-	60	20	-
		Rainfed	40	20	-	-	-	10	40	20	-
5.	Madhya Pradesh	Irrigated	60	30	20	60	30	20	-	-	-
		Rainfed	30	20	10	40	20	10	-	-	-
6.	Orissa	Irrigated	-	-	-	60	30	30	-	-	-
		Rainfed	30	15	15	30	15	15	-	-	-
7.	Punjab	Irrigated	100	50	-	60	40	-	50	20	-
		Rainfed	40	20	-	40	-	-	40	-	-
8.	Rajasthan	Irrigated	60	40	-	40	20	-	-	-	-
		Rainfed	30	20	-	20	10	-	-	-	-
9.	Uttar Pradesh and Uttaranchal	Irrigated	120	40	40	60	30	20	50	30	30
		Rainfed	60	20	20	40	20	-	30	20	20
10.	West Bengal	Irrigated	80	40	40	60	30	30	80	40	40
		Rain fed	50	25	25	40	20	20	40	20	20

Secondary nutrients

Sulphur (S): 30.0 kg/ha

Sulphur should be applied through Single Super phosphate and Gypsum.

Micronutrients

Boron (B)

Borax: 10.0 kg/ha

Molybdenum (M)

Ammonium molybdate: 0.5 kg/ha

If deficiency is apparent in soil test report or in previous crop, micronutrients should be applied once in 3–4 crops.

Method of application

Manures

Manures should be applied at the time of land preparation and well mixed with the soil by cross– ploughing followed by planking.

Fertilizers

- **Rainfed condition** - Total amount of nitrogen, phosphorus and potassium as basal dressing

- **Irrigated condition** -
  - **Basal dressing**: Total amount of phosphorus and half amount of nitrogen and potassium each
  - **Top dressing**: Remaining amount of nitrogen and potassium

20.8. INTERCULTURAL OPERATION

1. Weeding

Rapeseed and mustard field is heavily infested with some winter and perennial weeds (e.g., *Argemone maxicana, Cyperus rotundus, Chenopodium album, Melilotus alba, Cirsium arvense, Fumaria parviflora, Cynodon dactylon, Anagalis arvensis, Solanum nigram* etc). Weed causes approximately 20–30 per cent reduction of yield of the crop. The field should be kept free from weeds. One hoeing in case of toria in the third week after sowing and 1–2 hoeings in case of Mustard and sarson are adequate. The weeds near the crop plant should be removed by hand weeding. Weeds can also be controlled by using any one of the herbicides given in Table 124.

Table 124. Herbicides used for controlling the weeds in Mustard and rapeseed

Sr. No.	Names of herbicides	Rate (kg.ai.ha^{-1})	Time of application
1.	Fluchloralin 45 EC	1.0	Pre–plant or pre–emergence
2.	Pendimethaline 30 EC	1.25	Pre–plant or pre–emergence
3.	Fluzifop-p-butyl 12.5 EC	0.250	Post–emergence

2. Thinning

Thinning is done three weeks after sowing to maintain a plant to plant distance of 10–15 cm.

20.9. IRRIGATION

Mustard and rapeseed are grown as irrigated and rainfed crops. Flowering and pod filling stages are the most critical growth stages for water requirement. Irrigation should be given as per the requirement of the crop.

20.10. HARVESTING

Mustard and toria should be harvested when the leaves turn yellow and the pods are just ripe. Sarson can be harvested at comparatively dry ripe stage as it is less liable to shattering of seeds. Harvesting should be done with a hand sickle or by uprooting of plants. The harvested plants should be carried to the threshing floor and dried either in bundles keeping the roots downward or spreading the plants on the threshing floor for about 4–5 days before threshing.

Table 125. Maturity period for harvesting of Mustard and rapeseed

Sr. No.	Type of Mustard-Rapeseed	Period of maturity for harvesting
1.	Toria	70–100 d
2.	Indian mustard	105–160 d
3.	Yellow sarson	120–155 d
4.	Brown sarson	100–235 d
5.	Taramira	140–150 d

20.11. THRESHING

Threshing can be done by beating the dry plants with bamboo sticks or wooden mallets or by trampling under the feet of the bullock. Nowadays, threshing is done by using paddy thresher. The seeds can be cleaned by traditional method of winnowing. Winnowing should be done with the help of natural air current by slowly dropping the threshed produce from a basket held at shoulder height.

20.12. YIELD

The yield varies according to the variety. In the climatic conditions similar to that of West Bengal, the yield of different varieties is given in Table 126.

Table 126. Yield of Mustard and rapeseed

Sr. No.	Type of Mustard-Rapeseed	Variety	Yield (kg/ha)
1.	Toria	B-54	1,250.0
2.	Rai	B-85	1,750.0
		T-59	2,000.0
3.	White Sarson	B-9	1,500.0

20.13. STORAGE OF SEED

The cleaned seeds, after being dried in the Sun for 2–3 days or till the moisture content comes down to eight per cent, are stored in gunny bag or pusa bin. However, there should not be high humidity and moisture in the store house. The undried and unclean rapeseed deteriorates quickly in bulk. Dry unbroken rapeseed is generally free of storage pest than other oilseeds, the hard seed coat offering considerable protection.

20.14. VARIETIES

The varieties of Mustard and rapeseed recommended for cultivation in different States are as follows:

Table 127. Varieties of Mustard and rapeseed recommended for cultivation in different States

S.N.	State	Strain	Recommended varieties
1.	Assam	Yellow sarson	M-3, M-18 etc.
		Toria	BR-23, M-27, PT-303 etc.
2.	Bihar	Rai	BR-13, BR-40, Pusa Bold, T-59 (Varuna) etc.
		Toria	BR-23, BR-29, BR-32, BR-36 etc.
3.	Gujarat	Rai	Patan-67 etc.
		Yellow sarson	Patan sarson-66,YSPB-24 etc.
4.	Haryana	Rai	Prakash, RH-29, RH-30 etc.
		Brown sarson	BHS-1, Suphala etc.
		Toria	Sangam etc.
5.	Madhya Pradesh	Rai	Rohini,Vaibhav,Vardan etc.
		Toria	T-9, T-36 etc.
6.	Orissa	Rai	Bhagirathi (RW-351), Pusa Bold etc.
		Toria	M-27, TS-29 etc.
7.	Punjab	Rai	RI-18, RLM-198, RLM-514, RLM-619 etc.
		Toria	ISTA, TL-15 etc.
		Brown sarson	Pusa kalyani etc.
		Yellow sarson	YS-24 etc.
8.	Rajasthan	Rai	RL-18 etc.
		Toria	T-9,T-36 etc.
		Brown sarson	BSH-1, Pusa Kalyani etc.
9.	Uttar Pradesh and Uttaranchhal	Rai	T-16, T-59 (Varuna), RT-1, Laha-101, Shekhar, Kranti, Rohini, Vaibhav, Krishna, Narendra Rai-8501 etc.
		Toria	Bhabani, T-9, T-36, PT-303, PT-507 etc.
		Brown sarson	BS-2, Pusa Kalyani, RS-70 etc.
		Yellow	K-88, T-151, Type-42 etc.
10.	West Bengal	Rai	T-59 (Varuna), B-85 (Seeta), Bhagirathi (RW-351), Pusa Bold, Sarama (RW-85-59), Apprest Mutant, T-11, T-16, Pusa Bold, Sanjukta etc.
		Toria	B-54 (Agrani), Panchali (TWC-3), T-9, T-36, T-39, RL-8, RL-15, Tori-7 etc.
		White Sarson	B-9 (Binoy), Salek etc.
		Brown sarson	Pusa Kalyani, BS-1, BS-2, BS-70, DST-324 etc.

Table 128. Improved varieties of Mustard and rapeseed in India

Sr. No.	Type	Variety	Area of adaptability	Duration (days)	Suitable characteristics	Yield (q/ha)
1.	Toria varieties	T-36 (Yellow)	Uttar Pradesh	97–100	Medium tall, yellow seed, oil 44%	12–15
		Sangam	Haryana, Assam	100–110	High oil content 41%	12–15
		ITSA	Uttar Pradesh, Punjab, Haryana	100–110	Tall, profuse branching, oil 44%	8–10
		Bhavani	Uttar Pradesh	80–85	Oil 43%	12–13.5
		PT-507	Uttar Pradesh	90–95	Oil 42–43%	15–18
		TL-15	Uttar Pradesh	80–90	Oil 44%	10–12
		PT-303	Uttar Pradesh	90–95	Oil 42–44%	15–18
		PT-30	Uttar Pradesh	90–95	Oil 43.2%	14–16
		Gaurani	Uttar Pradesh	85–90	Oil 42.4%	12–15
		D.K-1	Himachal Pradesh	80-90	Early Variety, oil 44%	12–15
		T-9 (Black)	Uttar Pradesh, Madhya Pradesh, Bihar	90–95	Dwarf, oil 44%	12–15
2.	Yellow mustard	T-151	Eastern Uttar Pradesh	120–125	Oil 46%	14–15
		K-88	Central and western Uttar Pradesh	130–132	Oil 43%	14–18
		66-197-3	Bihar	120–125	Oil 42 %	14–16
		P.S-66	Northern Gujarat	110	Oil 46%	12–14
		Vinay (B-9)	West Bengal and eastern States	90–95	Oil 46%	12–14
3.	Rai	T-59 (Varuna)	Plains	125–130	Bold seed, oil 39.8%	20–25
		Kranthi	Plains	125–130	Powdery mildew and white rust resistant, oil 40%	20–25
		Krishna	Plains	128–132	Bold seed, frost resistant, oil 40.2%	22–28
		Rohini (KRB-24)	Whole Uttar Pradesh	130–135	Bold Seed, pest and disease resistant, oil 41.1%	22–28
		Vardan (Rk-1468)	Whole Uttar Pradesh	125–130	Suitable for late and mixed cropping, oil 40%	22–28

Narendra rai (8501)	Whole Uttar Pradesh	125	Suitable for alkaline and saline soils, oil 40%	25–30
Vaibhav (PK-1481)	Whole Uttar Pradesh	125–130	White rust resistant, oil 37.8%	15–20
RLM-198	Whole India	152–166	Excessive minerals and aphids resistant, oil 38%	10–12
Laha-101	Uttar Pradesh	145–150	Oil 42%	13–18
Pusa Bold	Delhi, eastern India	135–140	Bold seed, pods do not shatter, oil 40%	18–26
Prakash	Haryana	150–155	Oil 40%	15–20
Shekhar	Uttar Pradesh	120–125	Medium size plant, bold seed, oil 40%	
Seeta (B-85)	West Bengal, Andhra Pradesh	90–95	Oil 38%	10–12
Patan	Gujarat	119	Oil 38%	16–19
Baagi-rathi	West Bengal, Orissa	110	Oil 38%	12–14
RLM-514	West Bengal, Andhra Pradesh	150–52	Oil 40%	15–20

20.15. PLANT PROTECTION

Insect pests

1. Mustard aphid

Scientific Name - *Lipaphis erysimi* Davis

Syn. - *Hyadaphis erysimi*

The aphid is a serious and destructive insect pest of Mustard and rapeseed and it occurs in all growing tracts of India. The insect appears in the field towards the end of November and remains active till the end of February when the temperature starts rising.

Nature of damage

Both nymph and adults suck the sap from leaves, succulent tender stem, inflorescence or the developing pods. The heavily–infested plant remains dwarf and stunted and new branches are not formed. The petals of open flowers turn prematurely white, and the pod shrivels up and does not develop under severe infestation. Aphids also secrete liquid called honeydew on which black mould develops giving the plants a blighted appearance. The moulds adversely affect the normal physiological activities of the plant. In severe infestation, the plants may die. Cloudy and moist weather is favourable for the multiplication of aphid.

Fig 44. Mustard aphid–affected plant

Control measures

a. Cultural method

- The aphid–tolerant varieties such as JM-1 and RK-9501 should be cultivated.
- The crop should be sown earlier than normal sowing (i.e., before 20th October) as the earlier sown crop escapes the severity of aphid infestation.
- The recommended doses of fertilizers should be applied. Application of excessive amount of nitrogenous fertilizers should be avoided.

b. Mechanical method

The affected parts along with aphid population in the initial stage should be destroyed by burning or burrowing under the soil.

c. Biological method

- Ladybird beetles viz., *Cocciniella septempunctata, Hippodamia variegate* and *Cheilomones vicina* are most efficient predators of the mustard aphid. Adult beetle may feed an average of 10–15 adults/day.
- Several species of Sypris fly i.e., *Sphaerophoria* sp, *Eristallis* sp, *Metasyrphis* sp, *Xanthogramma* sp and *Syrphus* sp predate on aphid.
- The lacewing, *Chrysoperla carnea* predates on the mustard aphid colony.

d. Chemical method

The aphid can be controlled by spraying any one of the insecticides given in Table 129.

Table 129. Insecticides used for controlling Mustard aphid

Sr. No.	Name of insecticide	Amount (per litre of water)
1.	Dimethoate 30 EC	2.0 ml
2.	Acephate 75 WP	0.75 g
3.	Cartap hydrochloride 50 SP	1.0 g
4.	Clothianidin 50 WDG	0.1 g

2. Bihar hairy caterpillar (BHC)

Scientific name - *Spilosoma obliqua*

The Bihar hairy caterpillar is a polyphagous insect and causes severe damage to Mustard and rapeseed when it appears in epidemic form.

Nature of damage

BHC feed on leaves, young shoot and green pods, and in severe infestation the whole crop is defoliated. Young larvae remain in cluster on the lower surface of the leaves and feed gregariously, mostly on the under surface of the leaves. The grown–up caterpillars migrate from one field to other and feed on the leaves voraciously leaving behind only the midrib. When they attack at green pod stage, they feed on entire green seeds. As a result, the yield is decreased.

Control measures

a. Cultural method

- Pre–monsoon deep ploughing (2–3 times) should be given as this process exposes the hibernating pupae to sunlight and predatory birds.
- The alternate wild host, which harbours the hairy caterpillars, should be removed and destroyed by burning.

b. Mechanical method

The egg masses and early instar larvae should be collected and destroyed by burning or burrowing under the soil.

c. Biological method

The natural bio–control population of spiders, long–horned grasshoppers, praying mantid, robber fly, ants, green lacewing, damsel flies/dragon flies, flower bugs, shield bugs, lady bird beetles, ground beetle, predatory cricket, earwig, braconids, trichogrammatids, NPV (Nuclear Polyhydrosis Virus) and green muscular fungus should be used.

d. Chemical method

The Bihar hairy caterpillar can be controlled by spraying any one of the insecticides given in Table 130.

Table 130. Insecticides used for controlling Bihar hairy caterpillar of Mustard and rapeseed

Sr. No.	Name of fungicide	Amount (per litre of water)
1.	Acephate 75 WP	0.75 g
2.	Thiomethoxam 25 WG	0.2 g
3.	Lamdacyhalothin 5 EC	0.30 ml
4.	Novaliuron 10 EC	1.0 ml

3. Mustard saw fly

Scientific name - *Athalia lugens proxima* Klung

The Mustard saw fly is most serious insect pest of Mustard and rapeseed. This insect pest appears in the month of October and maximum damage is caused during the months of October–December. The female has saw–like ovipositor. Hence, it is called saw fly.

Nature of damage

The larvae are greenish black in colour and apparently look like caterpillar, which feed on leaves of young seedlings making holes, mostly in the morning or at dusk. The young larvae nibble leaves of affected plants and later it feeds from the margins toward the midrib. The affected leaves become curled and fall on the ground when touched. They may eat all the leaves in case of heavy attack. It is a sporadic insect and in certain years, it becomes so serious that the plants are defoliated completely. Older plants when attacked do not have seeds in the pod.

Control measures

a. Cultural method

- Summer ploughing is most effective to destroy the pupa.
- Early sowing should be done as the early–sown crop escapes from the attack of the insect.

b. Mechanical method

The grubs of saw fly may be collected in the morning or evening, and destroyed by burning or burrowing under the soil.

c. Biological method

- The bitter gourd seed emulsion may be used as anti–feedent.
- *Perilissus cingulator* and the bacterium *Serratia marcescens* Bizio may be conserved as they parasite the grubs and infect the larvae of sawfly respectively.

d. Chemical method

The insect pest can be controlled by spraying any one of the insecticides given in Table 131.

Table 131. Insecticides used for controlling Mustard saw fly

Sr. No.	Name of insecticide	Amount (per litre of water)
1.	Quinalphos 25 EC	2.0 ml
2.	Malathion 50 EC	2.0 ml
3.	Thiomethoxam 25 WG	0.2 g
4.	Spinosad 45 SC	0.25 ml

4. Painted bug

Scientific name - *Bargrada hilaris*

The painted bug is most serious insect pest of mustard and rapeseed. It is widely distributed across India. The insect is more serious when the seedlings are just emerging.

Nature of damage

Adults and nymphs suck sap from leaves and stems resulting in poor growth and pale yellow colour of the leaves. In the later stage, they suck the sap from the pod. As a result of which, the quality and quantity of yield is adversely affected. The young plants, when attacked by this insect, may wilt and afterwards die. The bugs also feed on mustard seeds.

Control measures

a. Cultural method

- Deep ploughing may be given so that the eggs of painted bug are destroyed.

- The crop should be sown early as the early–sown crop escapes the attack of this insect.

b. Mechanical method

The remains of mustard crop should be collected and destroyed by burning so that the insects of different stages do not reach the next year crop.

c. Biological method

The bio–control agents such as *Alophara* sp should be conserved.

d. Chemical method

The insect pest can be controlled by spraying any one of the insecticides given in Table 132.

Table 132. Insecticides used for controlling the Painted bug of Mustard and rapeseed

Sr. No.	Name of insecticide	Amount (per litre of water)
1.	Malathion 50 EC	2.0 ml
2.	Acephate 75 WP	0.75 g
3.	Cartap hydrochloride 50 Sp	1.0 g

5. Diamondback moth (DBM)

Scientific name - *Plutella xylostella*

Nature of damage

The diamond back moth feeds on the foliage and leaves may be eaten up completely. Later on, it also bores into pods and feeds on developing seeds.

Control measures

a. Mechanical method

- The larvae at gregarious stage should be collected at least twice a week and destroyed by burning or burrowing under the soil.
- The pheromone trap may be installed in the field to control the adult insect.

b. Biological method

The *Cotesia plutellae* and *Diadegma insulare* should be conserved as they are important parasitism for diamondback moth.

c. Chemical method

The insect pest can be controlled by spraying any one of the insecticides given in Table 133.

Table 133. Insecticides used for controlling the Diamond back moth of Mustard and rapeseed

Sr. No.	Name of insecticide	Amount (per litre of water)
1.	Malathion 50 EC	2.0 ml
2.	Spinosad 45 SC	0.25 ml
3.	Thiomethoxam 25 WG	0.2 g
4.	Lufeneuron 48 EC	1.5 ml

6. Mustard leaf miner (MLM)

Scientific name - *Phytomyza articornis* Meigen

Leaf miner has been reported from all Mustard and rapeseed growing areas, attacking mainly rai and sarson. The insect appears in the month of December and most of the damage is done in the month of February.

Nature of damage

The adult female causes initial damage by puncturing young leaves at numerous places with their ovipositors for feeding or laying eggs. The polyphagous greenish–yellow larvae eat away the green matter by making serpentine mines in the leaves as a result of which photosynthesis gets disturbed.

Control measures

a. Cultural method

The crop should be sown early as the early–sown crop escapes the attack of leaf miner.

b. Mechanical method

The insect–affected leaves at initial stage may be collected, if possible and destroyed by burning.

c. Chemical method

The insect can be controlled by spraying any one of the insecticides given in Table 134, as and when its appearance is noticed.

Table 134. Insecticides used for controlling the Mustard leaf miner

Sr. No.	Name of insecticide	Amount (per litre of water)
1.	Dimethoate 30 EC	2.0 ml
2.	Acephate 75 WP	0.75 g
3.	Malathion 50 EC	2.0 ml
4.	Thiomethoxam 25 WG	g

7. Cabbage butterfly

Scientific name - *Pieris brassicae* Linn

Cabbage butterfly causes damage in Mustard and rapeseed in Uttar Pradesh and West Bengal. The larvae of this butterfly are yellow in colour.

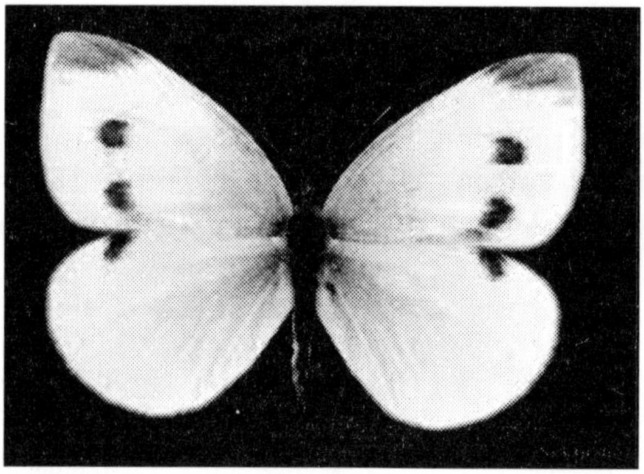

Fig 45. Cabbage butterfly (Adult)

Nature of damage

The young larvae feed gregariously on few plants in the field, but later they spread to other plants and infest the whole field. The grown–up caterpillars feed voraciously on leaves causing severe defoliation of the crop. They also feed on tender shoots, branches and green pods.

Control measures

a. Mechanical method

Hand picking of larvae in the initial stage is economical.

b. Chemical method

The insect can be controlled by spraying the crop with any one of the insecticides given in Table 135.

Table 135. Insecticides used for controlling the Cabbage butterfly of Mustard and rapeseed

Sr. No.	Name of insecticide	Amount (per litre of water)
1.	Dimethoate 30 EC	2.0 ml
2.	Acephate 75 WP	0.75 g
3.	Malathion 50 EC	2.0 ml
4.	Thiomethoxam 25 WG	0.2 g

Diseases

1. Alternaria blight

C.O.: *Alternaria brassicae*

The alternaria blight is a common and serious disease of Mustard and rapeseed. The disease occurs in the months of January–February.

Symptoms

Brown to blackish spots with concentric rings appear on leaves and later spread to stem and pod. The spots on the stem are oval shaped, but the spots on the leaves and pods become circular. In case of severe attack, the upper part of the stem withers. Defoliation also occurs in severe infection.

Control measures

As the disease is seed born, the seed should be treated with any one of the fungicides given in Table 136, before sowing, to minimize the disease infestation:

Table 136. Fungicides used for seed treatment for controlling the Alternaria blight of Mustard and rapeseed

Sr. No.	Name of fungicide/bio-fungicide	Amount (per litre of water)
1.	Carbendazim 50 WP	2.0–3.0 g
2.	M.E.M.C 6 WP	1.0 g
3.	Captan 50 WP	2.0–5.0 g
4.	Mancozeb 75 WP	2.5 g
5.	Trichoderma viridi	4.0–5.0 g

2. Sclerotinia stemrot

C.O.: *Sclerotinia sclerotiorum*

Symptoms

The water–soaked spots appear on the stem, which later may be covered with a cottony–white growth. As the disease progresses, affected portion of the stem develops a bleached appearance and eventually the tissues are shred. The girdling of stem results in pre–mature ripening of fruits and in lodging of plants.

Control measures

- Crop rotation should be followed, at least a five year rotation, particularly for severely–infested fields.
- The disease can be controlled by spraying the crop with any one of the fungicides given in Table 137.

Table 137. Fungicides used for controlling the Sclerotinia stem rot of Mustard and rapeseed

Sr. No.	Name of fungicide	Amount (per litre of water)
1.	Carbendazim 50 WP	1.0 g
2.	Captan 50 WP	2.5 g
3.	Copper oxychloride 50 WP	5.0 g

3. Downy mildew

C.O.: *Pernospora brassicae* Gaumann

Downy mildew is a serious disease of Mustard and rapeseed. In most of the cases downy mildew and white rust appear together. The disease occurs from December–January, mainly at the flowering stage.

Symptoms

All the aerial parts of the plants are attacked by the disease–producing organism, but the foliage (leaves) and seed–bearing part (inflorescence) are most severally affected. Thin, small, greyish white growth of downy mildew appears in patches on the lower surface of the leaves and the upper surface has marked white spots corresponding to patches below. Leaves dry up, shrivel and can be torn easily. The fungus also causes swelling and distortion of floral parts. No seeds are produced on the affected inflorescence.

Control measures

a. Cultural method

- Clean cultivation (i.e., control of weeds, destruction of diseased crop debris and residues, application of potash in recommended doses etc.) and crop rotation should be followed.

- The healthy seeds collected from disease–free field should be used for sowing and also they should be sown in time (i.e., between 10– 25th October).

b. Chemical method

The disease can be controlled by treating the seeds and spraying the crop with any one of the fungicides given in Table 138, at an interval of 10–15 days.

Table 138. Fungicides used for controlling Downey mildew of Mustard and rapeseed

Sr. No.	Name of fungicide	Amount (per litre of water)
1.	Mancozeb 75 WP	2.5 g
2.	Chlorothalonil 75 WP	2.0 g
3.	Dithiolone75 WP	2.5 g
4.	Propineb 70 WP	3.0 g

4. Bacterial rot

C.O.: *Xanthomonas campestris* pv. *campestris* (Pammel Dowson)

Symptoms

Dark–coloured streaks appear on the stem of the plant from ground level at 60 DAS, which gradually enlarge and girdle the stem. The stem of disease–infected plant becomes hollow due to internal rotting. The cracking of the midrib of lower leaves, browning of veins and withering is observed. In severe cases, the vesicular bundles of the stem also turn brown and the plant ultimately collapses.

Control measures

- The disease–affected plants should be uprooted and destroyed by burning.
- The disease can be controlled by spraying the crop with any one of the bactericides given in Table 139.

Table 139. Bactericides used for controlling the Bacterial rot of Mustard and rapeseed

Sr. No.	Name of bactericide	Amount (per litre of water)
1.	Streptocycline (9:1) SP	0.1 g
2.	Bacterimycin SC	0.4 g
3.	Validimycin 3 L	2.0 ml

5. Phyllody

C.O.: Mycoplasmal like organism (MLO)

The Phyllody is an important viral disease and it causes considerable damage on Toria and Yellow sarson in Punjab, Haryana and Uttar Pradesh. The disease is transmitted through the insect vector like leaf hopper.

Symptoms

The disease–affected parts of the plant are malformed. Numerous shoots arise and the plants become brushy or broom–shaped. The floral parts are transformed into leafy structure and such plants do not bear siliqua.

Control measures

- Early sowing should be avoided.
- The diseased plants should be roughed at early stages and the rest of the plants should be sprayed with any one of the insecticides given in Table 140 to control the insect vector.

Table 140. Insecticides used for controlling the insect-vector of Phyllody of Mustard and rapeseed

Sr. No.	Name of insecticide	Amount (per litre of water)
1.	Cartap hydrochloride 50 SP	1.0 g
2.	Acephate 75 WP	0.75 g
3.	Thiomiton 25 EC	2.0 ml

6. White rust

C.O.: *Albugo candida* (Pers.) Kuntz

Moist (more than 70 per cent relative humidity) and warm weather (12°C–25°C), and intermittent rains favour the disease development.

Symptoms

White or creamy–yellow scattered pustules of variable size and shape appear chiefly on the lower surface of the leaves. The pustules coalesce and form patches, which cover the entire leaf causing thickening and rolling of leaves. Sometimes stems and flowers are hypertrophied into various types of swellings. Flowers get malformed and sterile. All parts of the plants except root are affected.

Fig 46. Rust disease

Control measures

a. Cultural method

- The diseased debris of the previous crop should be destroyed by burning.
- The healthy certified seeds, after treating with seed treatment chemicals, should be sown during the period of15th September–15th October.
- Clean cultivation (i.e., control of weeds, removal of diseased plants etc.) and crop rotation should be followed.

b. Chemical method

The disease can be controlled by spraying the crop with any one of the fungicides given in Table 141.

Table 141. Fungicides for controlling the White rust of Mustard and rapeseed

Sr. No.	Name of fungicide	Amount (per litre of water)
1.	Tridemorph 80 EC	1.0 ml
2.	Carbendazim 50 WP	1.0 g
3.	Metalaxyl 35 WS	0.5 g
4.	Flutrazol 5 SC	2.0 g

21 BAY LEAF

Scientific name - *Laurus nobilis L*

Family - Lauraceae

Commercial part - Leaf

Indian names - *Bengali* : Tejpata; *Hindi* : Tejpatta; *Assamese* : Mahpat, Tejpat; *Tamil* : Talishappattiri; *Gujarati* : Tamaal patra; *Marathi* : Tamaal patra; *Malayalam* : Tamalapatram; *Urdu* : Tezpat; *Manipuri* : Tejpat; *Telugu* : Talisapatri, Talisha, Patta akulu; *Sanskrit* : Tamalapattra.

Bay leaf is one of the oldest herbs or spice. It is native to Asia Minor and is known by the name of sweet bay, bay laurel, noble laurel and true laurel. The leaves have an aromatic odour and are bitter and spicy. Bay leaves refer to the aromatic leaves of several plants used in cooking. The upper surface of the leaf is glabrous and shiny, olive green in colour, and lower surface is dull olive to brown in colour with veins and a prominent rib.

Fig 47. Bay leaf

21.1. ORIGIN AND DISTRIBUTION

Bay leaf is native to Asia Minor from where it reached Mediterranean. It is also grown in the regions of France, Belgium, central and north America, Turkey, Italy, Russia Israel and India. It is widely cultivated in Europe, America and Arabian countries. It is not cultivated as a commercial crop in India.

21.2. CLIMATE

Bay trees grow best in warm, moist climate, similar to that of the Mediterranean area. Bay trees need plenty of Sun to grow well. Bay laurels thrive in at least partial Sun. The amount of exposure of sunlight should be at least 6-8 hours per day, and preferably eight. This plant cannot stand extremely cold weather conditions, and requires a warm climate for its healthy growth.

21.3. SOIL

The trees prefer well–drained soil that is rich in nutrients. Bay is not too particular about soil and can tolerate soil pH from 4.5–8.3.

21.4. PROPAGATION

There are three methods of propagating:

1. Seed propagation

Bay leaves can be propagated by seeds. For raising seedlings, well–ripe fruits should be collected from high–yielding plants. The seeds should be sown either in nursery or polythene bags filled with a mixture of 1:1:1 sand, soil and FYM. The crop can be raised as a pure or as a perennial intercrop of coconut, arecanut etc. Pits should be of the size 75 x 75 x 75 cm in hard and laterite soils, and 50 x 50 x 50 cm in sandy and alluvial soils. The spacing should be 4 x 4 m for grafting and 7 x 7 m for seedling. Before planting, the pits should be filled with top soil and 2 kg of compost or well–decomposed cattle manure, and 10 g of Carbyl dust to avoid white ants attack. Generally, planting should be done during the months of July–October.

2. Vegetative propagation

Cutting

Late summer and early autumn are the best times to propagate this tricky herb. A young side–shoot about 10–15 cm long should be stripped from the parent plant, and lower leaves and any trilling strands should be trimmed off. Cuttings should be about 4 inches long and treated with a rooting hormone. The cuttings should be placed in a container filled with a porous rooting media.

Layering

Layering is also a way to propagate this plant. Bend a branch gently until it lays on ground. A nick is made in the branch where it touches the ground and it is hold in place against the soil with a rock on either side or using garden staples. The hormone powder may be used on the nicked area for rooting promptly.

21.5. PLANTING

The soil should be brought to a fine tilth by ploughing with tractor and power tiller followed by planking. Planting in the main field should be done during the monsoon season. Pit size of 30 x 30 x 30 cm should be dug one to two months before planting at a spacing of 1.2–1.5 m. One year old seedlings are suitable for planting. One seedling should be planted at the centre of the pit in the main field. It can also be propagated by root suckers. There are a number of root suckers near its plants. They are separated from the main plant during rainy season and planted immediately in the main field. The main field should be ploughed thoroughly. FYM @ 10 kg/pit should be applied.

21.6. MANURING

The amounts of manures and fertilizers that are to be applied per plant per year are as follows:

Manures

Compost or FYM: 10.0 kg

Fertilizers

Nitrogen (N)	: 60 g
Phosphorus (P_2O_5)	: 80 g
Potassium (K_2O)	: 40 g

The fertilizers may be applied at the onset of the monsoon.

21.7. INTERCULTURAL OPERATION

1. Weeding

The field of Bay leaf should be kept free from weeds. For this, weeding should be done to control weeds. Pruning is the biggest part of maintenance of plant. Bay is often kept pruned, either to keep the size in check or to attain a more early growth stage of crop.

2. Training and pruning

The plants should be trained and pruned to maintain a height of 1 m. Pruning is usually done in the spring, when new growth just starts.

21.8. IRRIGATION

Irrigation should be given immediately after planting. The second irrigation should be given on the third day and then the irrigation should be given once in a week.

21.9. HARVESTING

The bay is harvested once the plant is a couple of years old. Bay leaves can be harvested year round (as it is an evergreen tree). The leaves should be dried before use, as fresh bay leaves are bitter.

The older the leaf, the stronger is the flavour. The leaves are best used in cooking after having been properly dried for at least a few days. One way to dry the leaves is to hang a branch upside down in a dry location for a week or so.

The composition of dried bay leaves (100 g) is as follows:

Table 142. Composition of dried bay leaves per 100 g

Constituents	Amount
Moisture (g)	4.5
Protein (g)	7.6
Fat (g)	8.8
Fibre (g)	25.2
Carbohydrates (g)	50.2
Total ash (g)	0.7
Calcium (g)	1.0
Phosphorus (g)	0.11
Sodium (g)	0.02
Potassium (g)	0.6
Iron (g)	0.53
Vitamin B1 (Thiamine) (g)	0.10
Vitamin B2 (Riboflavin) (mg)	0.42
Niacin (mg)	2.0
Vitamin C (Ascorbic acid) (mg)	46.6
Vitamin A (International units - I.U.)	545
Calorific value (food energy)	410

21.10. YIELD

Dry leaves - 13 kg per plant (Parajuli 1998)

21.11. USES

- Bay leaf is commonly used as a whole in stews, sauces, soups, fish, meat, and in confectionaries for enhancing their flavour.
- Bay leaf can also be used as a weevil deterrent.
- Bay leaf is a pungent addition to potpourri. An ointment made from Bay leaf can help reduce joint inflammation. The major functional properties are anti–microbial, anti–fungal, hypoglycaemic, anti–ulcerogenic etc.

22 SAFFRON

Scientific name - *Crocus sativus*

Family - Iridaceae

Commercial part - Flower

Indian names - *Bengali* : Jafran; *Hindi* : Kesar; *Gujarati* : Keshar; *Kannada* : Kumnkuma kesari; *Kashmiri* : Kong; *Malayalam* : Kunkumapoove; *Marathi* : Keshar, Kesara; *Punjabi* : Kesar, Zafran; *Sanskrit* : Keshara, Kunkuma, Aruna, Asra, Asrika; *Tamil* : Kungumapoo; *Urdu* : Zafran.

Saffron is the most precious and most expensive spice in the world. Saffron is a spice derived from the flower of *Crocus sativus*, commonly known as the saffron crocus. The flower is collected and that flower produces the spice. Saffron (*Crocus sativus* var. kashmiriana) is a very important, high value, low volume condiment and a cash crop of Kashmir. Saffron is characterized of having a purple flower with red stigmas and yellow stamens. The plant is a native of Iran and Asia Minor. Saffron is the only spice that comes from a flower. Saffron is a perennial, low–growing herb with a globular corm ranging from 0.5–5 cm in diameter. Saffron is a spice obtained from the stigmas of the flower of *Crocus sativus Linnaeus*, commonly known as 'Rose of saffron'.

Table 143. Nutritive value of Saffron

Nutritional value per 100 g (3.5 oz)	
Energy	1,298 kJ (310 kcal)
Carbohydrates	65.37 g
- Dietary fibre	3.9 g
Fat	5.85 g
- Saturated	1.586 g
- Monounsaturated	0.429 g

- Polyunsaturated	2.067 g
Protein	11.43 g
Water	11.90 g
Vitamin A	530 IU
Thiamine (vit. B_1)	0.115 mg (10%)
Riboflavin (vit. B_2)	0.267 mg (22%)
Niacin (vit. B_3)	1.460 mg (10%)
Vitamin C	80.8 mg (97%)
Calcium	111 mg (11%)
Iron	11.10 mg (85%)
Magnesium	264 mg (74%)
Phosphorus	252 mg (36%)
Potassium	1724 mg (37%)
Sodium	148 mg (10%)
Zinc	1.09 mg (11%)
Selenium	5.6 μg
Folate	93 μg
Vitamin B_6	1.010 mg
Ash	5.45 g

Only thread portion is edible.
Percentages are relative to US recommendations for adults.
Source: USDA Nutrient Database

22.1. ORIGIN AND DISTRIBUTION

Saffron is a native of southern Europe and cultivated in Mediterranean countries, particularly in Spain, Austria, France, Greece, England, Turkey, and Iran. In India, it is cultivated in Jammu and Kashmir and in Himachal Pradesh.

22.2. CLIMATE

Saffron thrives well in sub–temperate regions ranging from 1,500–2,400 m. It grows well at a day temperature of 20°C–22°C. An optimum of 12 hours light duration is essential for its growth and flowering. It grows well in an area that receives 30–40 cm rainfall. Dry weather condition during flowering period is essential for getting higher yields. The cultivation of Saffron needs an extreme climate; hot and dry weather in summer and cold in winter.

22.3. SOIL

Saffron can be grown in a wide range of soils characterized by moderate water–holding capacity and good water infiltration. Saffron grows well in drained loamy soil rich in calcium and high in organic matter. The soils should be deep and free from stones. The optimal soil pH value should be 7.0–7.5.

Fig 48. Saffron

22.4. LAND PREPARATION

Land should be brought to fine tilth by shallow ploughing 4–5 times with tractor or power tiller giving sufficient intervals between two ploughings. Land should be levelled and divided into small plots and channels to facilitate proper irrigation and intercultural operations. The manures should be applied at the time of land preparation and mixed well with soil by ploughing followed by planking.

22.5. PROPAGATION

Saffron is propagated by planting the corms. The mother corm reproduces annually and gives rise to four to six daughter cormlets.

- **Time of planting :** June and July
- **Seed rate :** 1,600–2,000 kg/ha

- **Seed treatment**

The corms should be treated with any of the fungicides given in Table 144, before planting.

Table 144. Fungicides used for treatment of Saffron corms

Sr. No.	Name of fungicide	Amount (per kg of corm)
1.	Carbendazim 50 WP	2.5 g
2.	Captan 50 WP	3.5 g
3.	Trichoderma viridi	4.0 g
4.	Thiram 75 WP	2.5 g

Method of planting

The corms should be planted at a depth of 15–20 cm with a row to row spacing of 15 cm and corm to corm spacing of 5–8 cm. The corms can be planted by any one of the following methods:

- Ridge planting method
- Flat bed planting method
- Traditional planting method

22.6. MANURING

Manures: Compost or FYM: 20.0 t/ha

Fertilizers

Nitrogen (N)	: 50.0 kg/ha
Phosphorus (P_2O_5)	: 40.0 kg/ha
Potassium (K_2O)	: 20.0 kg/ha

22.7. HARVESTING

The flowering season is confined to three weeks from middle of October to first week of November. The flowers should be picked daily in the morning, and stigmas and styles should be trimmed immediately. About 1,60,000 flowers need to be handpicked to produce 1 kg of good quality dried Saffron. Saffron is harvested from the stigma lobes of the fall–blooming '*Crocus sativus*'.

22.8. USES

General uses

- Saffron is used as a culinary seasoning and for colouring different food materials. It has a unique taste that adds a distinctive flavour to foods.
- Saffron is an ingredient used for cosmetics and perfume industries.
- Saffron is used as dye for textile production.

Medical uses

- Saffron is used in fevers, melancholia, and for enlargement of liver and spleen. In Ayurvedic medicine, it is used to heal arthritis, impotency and infertility.
- Saffron has anti–cancer properties and suggests a positive healing effect on cancer patients.

22.9. PLANT PROTECTION

Insect pests

Saffron faces low insect pests as it spends most of its growth period in winter at low temperature.

Diseases

Corm rot

C.O.: *Fusarium oxysporium* f. species *gladioli*

Corm rot is a major disease of Saffron and it is a soil and seed–born disease.

Symptoms

The disease is characterized by yellowing and wilting of leaves due to basal stem rot and development of black round spot on the corm.

Control measures

- Summer ploughing helps in killing the resting spores of pathogen.
- The healthy and disease–free corms should be used for planting and planting should be done in raised bed with proper drainage.
- Field sanitation (i.e., collection and destruction of disease–affected plant parts including corm, and control of weeds etc.) should be followed.
- The corm should be treated with any one of the seed treatment materials given in Table 145.

Table 145. Seed treatment materials for controlling the Corm rot of Saffron

Sr. No.	Seed treatment material	Amount (per kg of corm)
1.	Carbendazim 50 WP	2.0–3.0 g
2.	Trichoderma viridi	4.0 g

- Application of Trichoderma mixed with FYM is a good method of reducing the incidence of corm rot.

23 VANILLA

Scientific name - *Vanilla planifolia*

Syn. -*Vanilla fragrans*

Family - Orchidaceae

Commercial part - Fruit

Common and other names - Vanillin, vainilla meaning *little sheath,* vanille, vaniglia, wanila, vanilka, vanilje, vanilja, vanil, vanillia, paneli, vanile, vanilya

Vanilla is a flavour derived from orchids of the genus *Vanilla*, primarily from the Mexican species, flat–leaved vanilla (*Vanilla planifolia*). The word 'vanilla', derived from the diminutive of the Spanish word *vaina* (*vaina* itself meaning sheath or pod), simply translates as little pod. *Vanilla planifolia* is a species of vanilla orchid. It is native to Mexico, and is one of the primary sources for vanilla flavouring, due to its high vanillin content.

23.1. ORIGIN AND DISTRIBUTION

Vanilla is native to Mexico but, now it is cultivated throughout the tropics. Main countries producing Vanilla include Madagascar, Mexico, Tahiti, Malagasy Republic, Comoro, Reunion, Indonesia, Seychelles and India. Madagascar so far is the largest producer of Vanilla in the world. India, China and Indonesia are the major vanilla–producing countries in Asia.

23.2. TYPES OF VANILLA

- **Bourbon vanilla** - Bourbon vanilla is typically grown in Indian Ocean islands like Madagascar, the Comoros, and Reunion.
- **Mexican vanilla** - Mexican vanilla comes from the place of its origin - Mexico.

- **Tahitian vanilla** - Tahitian vanilla originates from French Polynesia.

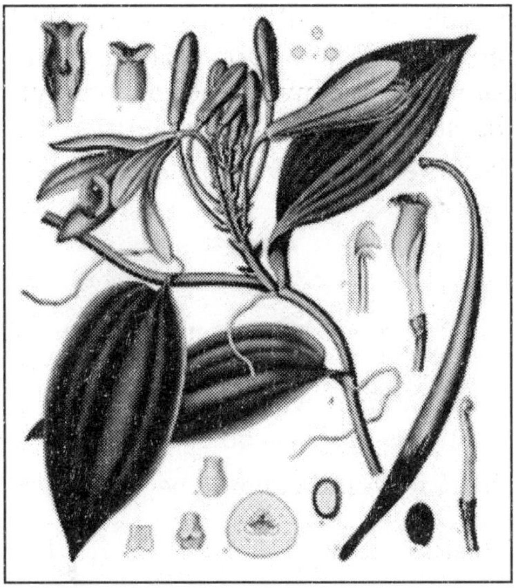

Fig 49. Vanilla

23.3. BOTANICAL DESCRIPTION

- **Plant** - The plant is climbing monocot possessing a stout, succulent stem. The stem is fleshy from which root is sprouted that clings to the trees parasitically.
- **Leaf** - The leaves are short petioled and oblong, and about 20 cm long.
- **Flower** - The flowers are yellowish green or white in colour, 6 cm long, 2.5 cm wide and grow in bunches.
- **Fruit** - The fruit that is also known by the name of 'beans or pod' is a capsule, nearly cylindrical in shape and is around 20 cm in length.

23.4. CLIMATE

Vanilla thrives well in warm and moist climatic conditions with well-distributed annual rainfall of 150–300 cm and a temperature ranging between 25°C–32°C.

Dry period of about two months is needed to restrict vegetative growth and induce flowering.

23.5. SOIL

Vanilla grows in any type of soil, but it grows well in sandy loam or laterite soils. It grows best in light, porous and friable soils with pH 6–7. Partial shade is essential for successful cultivation.

23.6. LAND PREPARATION

Land should be brought to fine tilth by shallow ploughing 4–5 times with tractor or power tiller giving sufficient intervals between two ploughings. Land should be levelled and divided into small plots and channels to facilitate proper irrigation and intercultural operations. The manures should be applied at the time of land preparation and are mixed well with soil by ploughing followed by planking.

23.7. PROPAGATION

Vanilla is generally propagated by stem cutting. The cutting with five internodes and 60 cm length should be used for planting. Roots start growing from the cuttings after four to eight weeks and show the initial signs of growth. Tissue cultured vanilla plantlets can also be used for field planting.

Depending on soil fertility, vanilla should be planted in rows 1.5–2.5 m apart and 2.5– 3.0 m in the rows.

The ideal time to plant vanilla in south Indian conditions is August–September when the intensity of the south west monsoon is low.

23.8. MANURING

Manures: Compost or FYM: 10 kg per plant

Fertilizers: The NPK requirement per plant is as follows:

Nitrogen (N)	: 40–60 g
Phosphorus (P_2O_5)	: 20–30 g
Potassium (K_2O)	: 60–100 g

23.9. INTERCULTURAL OPERATION

1. Weeding

The land should be kept free from weeds. Regular weeding should be done to keep the crop free from weeds.

2. Mulching

Mulching is a good method for conserving both moisture and soil. Dried organic matter, leaves, weeds, coconut leaves/ husks etc. can be used for mulching. The mulching materials add nutrient to the soil after their decomposition.

3. Shade management

Vanilla is a climbing vine. Hence, it requires support system. The support can come from either tree or other leguminous plants. The trees such as Jatropa,

Glyricidia, Casuarina sp etc. are used as support plants. June– September are the most suitable months for its cultivation. The plantation requires shade either partially or completely during the initial growth periods. The thumb rule is that vanilla requires about 50 per cent shade.

4. Trailing of vines

Trailing of vines is an important cultural operation. The growth of plants should be limited up to about 150 cm height. This is mainly to facilitate easy manual pollination. Pollination of the flower is usually done by hand. The vines should be trailed by coiling them around the branches of the support trees or on horizontal supports without letting them to touch the ground.

23.10. IRRIGATION

Irrigation should be given as per the requirement of the plant. Mist and sprinkler irrigation are more effective and good for growth of vines.

23.11. HARVESTING

After six weeks of fertilization, pods attain full size, but they mature only after 4–10 months. Vanilla beans are harvested when the fruit is ripe. The vanilla beans turn yellow on ripening.

23.12. YIELD

Cured beans - 300–400 kg/ha

23.13. USES

Vanillin is mainly responsible for the fragrance, flavour and aroma of vanilla essence. As a flavouring agent, it is used in the preparation of ice creams, chocolates, cakes, pastries, puddings, soft drinks, pharmaceuticals, liquors, perfumery and in nutraceuticals.

23.14. PLANT PROTECTION

Insect pest

1. Vanilla bug

Scientific name - *Hayomorpha picus* sp

Vanilla bug is a common insect of vanilla.

Nature of damage

The insect sucks the sap from vegetative bud and inflorescence, which leads to mottling.

Control measures

- The egg masses should be collected at early stage and destroyed by burning.
- The insect can be controlled by spraying the crop with any one of the insecticides given in Table 146.

Table 146. Insecticides used for controlling the Vanilla bug

Sr. No.	Name of insecticide	Amount (per litre of water)
1.	Acephate 75 WP	0.75 g
2.	Cartap hydrochloride 50 SP	1.0 g
3.	Chlothianidin 50 WG	0.1 g
4.	Thiometon 25 EC	2.0 ml

2. Vanilla vine weevil

Scientific name - *Sipalus* sp

Nature of damage

The black–coloured weevil bores into and eats the tender stem and leaf.

Control measures

- The weevil at adult stage should be collected and destroyed by burning or burrowing under the soil.
- The insect can be controlled by spraying the crop with any one of the insecticides given in Table 147.

Table 147. Insecticides used for controlling the Vanilla vine weevil

Sr. No.	Name of insecticide	Amount (per litre of water)
1.	Acephate 75 WP	0.75 g
2.	Cartap hydrochloride 50SP	1.0 g
3.	Chlothianidin 50 WG	0.1 g
4.	Thiometon 25 EC	2.0 ml

3. Caterpillar

Scientific name - *Plasia aurifera*

Nature of damage

The light–greenish caterpillar feeds on vegetative shoot and first leaf forming web.

Control measures

- Hand picking and destroying the caterpillar is the best method of controlling this insect.
- The insect can be controlled by spraying the crop with any one of the insecticides given in Table 148.

Table 148. Insecticides used for controlling the Caterpillar

Sr. No.	Name of insecticide	Amount (per litre of water)
1.	Acephate 75 WP	0.75 g
2.	Cartap hydrochloride 50SP	1.0 g
3.	Chlothianidin 50 WG	0.1 g
4.	Thiometon 25 EC	2.0 ml

4. White grub

Scientific name - *Scaarabaeldae* sp

Nature of damage

The grubs damage the roots of vanilla, which are mainly confined to the surface layer of the soil.

In case of severe infestation, the entire plant dies.

Control measures

- Summer ploughing is a good practice for controlling the white grub as it exposes the grubs present in the soil to the predators. The light trap is found good for controlling this insect.
- Soil drenching with 0.1 per cent Acephate 75 WP is a good method of controlling this insect.

5. Achatina

Scientific name - *Achatina fulica Bowdich*

Nature of damage

The insect feeds on young shoots and leaves. The affected parts turn whitish in colour.

Control measures

The insect can be controlled by spraying the crop with any one of the insecticides given in Table 149.

Table 149. Insecticides used for controlling Achatina

Sr. No.	Name of insecticide	Amount (per litre of water)
1.	Acephate 75 WP	0.75 g
2.	Cartap hydrochloride 50 SP	1.0 g
3.	Chlothianidin 50 WG	0.1 g
4.	Thiometon 25 EC	2.0 ml

Diseases

1. Root rot

C.O.: *Fusarium* sp, *Sclerotium* sp

Root rot disease causes serious damage to many vanilla plantations.

Symptoms

The leaves and stems become yellow in colour. The disease–affected leaves becomes flaccid, shrivelled and give a drooping appearance. Browning of roots is also observed, which leads to rotting and decaying.

Control measures

- The affected roots should be removed and destroyed by burning, if possible, or burrowing under the soil. Mulching should be reduced.
- The disease can be controlled by drenching the soil and spraying the crop with any one of the fungicides given in Table 150.

Table 150. Fungicides used for controlling the Root rot of vanilla

Sr. No.	Names of fungicides	Amount (per litre of water)
1.	Copper oxychloride 50 WP	5.0 g
2.	Captan 50 WP	2.5 g
3.	Carbendazim 50 WP	1.0 g
4.	Copper hydroxide 77 WP	2.0 g

2. Bacterial softrot disease

C.O.: *Erwinia carotovora pv.Carotovora*

Symptoms

This disease infects leaf and shoot, and shows soft rot symptoms.

Control measures

The disease can be controlled by spraying the crop with any one of the bactericide given in Table 151.

Table 151. Bactericides used for controlling the Bacterial soft rot of vanilla

Sr. No.	Names of bactericide	Amount (per litre of water)
1.	Streptocyline (9:1) SP	0.1 g
2.	Validmycin 3 L	2.0 ml
3.	Bacterimycin SC	0.3g

3. Stem blight

C.O.: *Phytophthora* sp

Symptoms

The brown spots appear on the leaves and stems followed by yellowing of the leaves, and rotting of the leaves and stems.

Control measures

The disease can be controlled by spraying any one of the fungicides given in Table 152 at an interval of 10–15 d as prophylactic spray.

Table 152. Fungicides used for controlling the Stem blight

Sr. No.	Name of fungicide	Amount (per litre of water)
1.	Copper oxychloride 50 WP	5.0 g
2.	Captan 50 WP	2.5 g
3.	Carbendazim 50 WP	1.0 g
4.	Copper hydroxide 77 WP	2.0 g

4. Stem rot

C.O.: *Fusarium oxysporum* and *Fusarium solani*

Symptoms

Water–soaked lesions appear on the stem, which gives the stem a brown–coloured appearance, resulting in rotting of tissues, leaves turning yellowish and drying off.

Control measures

- The disease–affected portion should be removed and destroyed by burning or burrowing under the soil.
- The disease can be controlled by spraying any one of the fungicides given in Table 153 at an interval of 10–15 days as prophylactic spray.

Table 153. Fungicides used for controlling the Stem rot of Vanilla

Sr. No.	Name of fungicide	Amount (per litre of water)
1.	Copper oxychloride 50 WP	5.0 g
2.	Captan 50 WP	2.5 g
3.	Carbendazim 50 WP	1.0 g
4.	Copper hydroxide 77 WP	2.0 g

5. Bean rot

C.O.: *Phytophthora* sp

Symptoms

Premature rotting and shedding of beans

Control measures

- The disease–affected beans should be collected and destroyed by burning, if possible, or by burrowing under the soil.
- The disease can be controlled by spraying any one of the fungicides given in the Table 154 at an interval of 10–15 days as prophylactic spray.

Table 154. Fungicides used for controlling the Bean rot of vanilla

Sr. No.	Names of fungicides	Amount (per litre of water)
1.	Copper oxychloride 50 WP	5.0 g
2.	Captan 50 WP	2.5 g
3.	Carbendazim 50 WP	1.0 g
4.	Copper hydroxide 77 WP	2.0 g

6. Shoot tip rot

C.O.: *Fusarium sclerotium* Sp

Symptoms

The shoot tip and tip of the leaves decay and that causes the loss of yield.

Control measures

- The disease–affected beans should be collected and destroyed by burning, if possible, or by burrowing under the soil.
- The disease can be controlled by spraying any one of the fungicides given in Table 155 at an interval of 10–15 days as prophylactic spray:

Table 155. Fungicide used for controlling the Shoot tip rot of Vanilla

Sr. No.	Name of fungicide	Amount (per litre of water)
1.	Copper oxychloride 50 WP	5.0 g
2.	Captan 50 WP	2.5 g
3.	Carbendazim 50 WP	1.0 g
4.	Chlothalonil 75 WP	2.0 g

7. Anthracnose

C.O.: *Colletotrichum gloeosporiodes* Penz.

The diseased organism attacks the stem apex, leaves and roots, resulting in wilting and falling of the fruits.

Symptoms

Small brown spots appear on the leaves, which eventually turn into bigger spots. It is prevalent during the monsoon season.

Control measures

- The disease affected leaves should be collected and destroyed by burning, if possible or by burrowing under the soil.
- Emericella nidulus was proven to be antagonistic to colletotrichum gloeosporiodes (Penz) causing anthraenose of vanilla panifolia.
- The disease can be controlled by spraying any one of the fungicides given in Table 156 at an interval of 10-15 days as prophylactic spray

Table 156. Fungicides used for controlling anthracnose of vanilla

Sr. No.	Name of fungicides	Amount (per litre of water)
1	Dodain 65 WP	0.5 g
2	Thiophanet methyl 70 WP	1.0 g
3	Dithiolone 75 WP	2.5 g

8. Vanilla mosaic

C.O.: *Vanilla mosaic virus*

Symptoms

The symptoms include mosaic pattern or mottling on foliage, stunted growth, sterility and leaf distortion.

Control measures

- The disease–affected plants should be uprooted and destroyed by burning or burrowing under the soil.
- The insect vector should be controlled by spraying any one of the insecticides given in Table 157.

Table 157. Insecticides used for controlling the insect-vector of Vanilla mosaic

Sr. No.	Names of insecticides	Amount (per litre of water)
1.	Acephate 75 WP	0.75 g
2.	Cartap hydrochloride 50 SP	1.0 g
3.	Chlothianidin 50 WG	0.1g
4.	Thiometon 25 EC	2.0 ml

24 ASAFOETIDA

Scientific name - *Ferula assa-foetida*

Family - Apiaceae

Commercial part - Latex

Indian names - *Hindi* : Hing; *Bengali* : Hing; *Gujarati* : Hing; *Kannada* : Hinger, Ingu; *Kashmiri* : Yang, Sap; *Malayalam* : Kayam; *Marathi* : Hinga; *Oriya* : Hengu; *Punjabi* : Hing; *Sanskrit* : Badhika, Agudagandhu; *Tamil* : Perungayam; *Telugu* : Inguva, Ingumo; *Urdu* : Hing.

Asafoetida is an herbaceous perennial plant growing up to 2 m tall, with stout, hollow, somewhat succulent stems with 5–8 cm diameter at the base of the plant. It has finely–toothed leaves, clusters of many white or yellow flowers in large compound umbels and a hollow stem growing from a fleshy taproot. It is the root that produces the spice. Asafoetida is the dried latex (gum oleoresin) exuded from the living underground rhizome or tap root of several species of Ferula (three of which grow in India).

24.1. ORIGIN AND DISTRIBUTION

Asafoetida is said to have originated in Rome and despite not being a native of India, it has been used in Indian medicine and cookery for ages. The species are distributed from the Mediterranean region to Central Asia. In India, it is grown in Kashmir and in some parts of Punjab. Therefore, it is possible to plan cultivation of this plant not only in Kashmir but also in Himachal Pradesh, Uttaranchal, north Bengal, Sikkim, Arunachal Pradesh and other areas having identical environmental situation. India is the major consumer of this spice.

The major supply of Asafoetida to India is from Afghanistan and Iran.

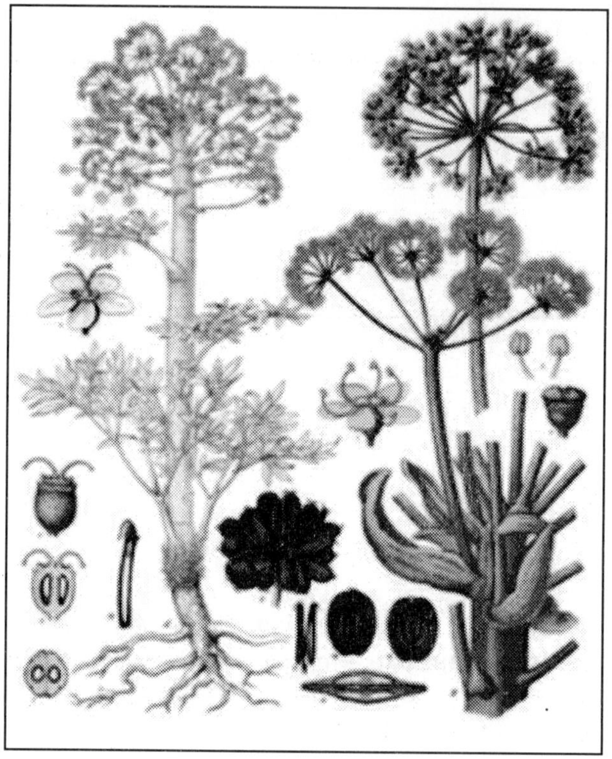

Fig 50. Asafoetida

24.2. SOIL

Asafoetida tolerates most soil types provided it has good drainage.

24.3. PLANTING

The seedlings, raised in a nursery bed, should be transplanted in outdoors in the spring at a spacing of 2 ft (60 cm) in sunny location when the danger of frost has passed.

24.4. VARIETIES

There are two main varieties of Asafoetida:

i. Hing kabuli sufaid (Milky white asafoetida)

ii. Hing lal (Red asafoetida)

Asafoetida is acrid and bitter in taste, and emits a strong disagreeable pungent odour due to the presence of sulphur compounds therein. The white pale variety is water soluble, whereas the dark or black variety is oil soluble.

24.5. HARVESTING

The time to start harvesting the gum from the rather succulent stem and the root is just before flowering, which is in the months of March and April.

An incision should be made in the upper part of the root or lower part of the stem and the exuding gum or latex should be collected.

24.6. USES

Asafoetida is extensively used for flavouring curries, sauces, and pickles. It is also used in medicines because of its antibiotic properties. This spice is used as a digestive aid and it also reduces flatulence.

25 CAPSICUM

Scientific name - *Capsicum annum* L., *Capsicum frutescens* L.

Family - Solanaceae

Commercial part - Green as well as ripe and dried pod (fruit)

Indian names - *Hindi* : Simla mirch; *Bengali* : Lanka, Lankamorich; *Gujarati* : Marcha; *Kannada* : Mensina kai; *Malayalam* : Mulaku; *Marathi* : Simla Mirchi; *Oriya* : Lanka; *Punjabi* : Lalmirch; *Tamil* : Milagay; *Telugu* : Mirapa kaya; *Urdu* : Lalmirch.

Capsicum is one of the important crops of India. It is used as vegetable and spice crop, and commonly known as 'Sweet pepper' or 'Simla mirch'. Green peppers are fleshy, blocky and in various shapes, more like a bell, hence named 'bell pepper'. The fruits of Capsicum are berry botanically.

Table 158. Nutritive value of Capsicum

Pepper, sweet, green raw - Nutritional value per 100 g (3.5 oz)	
Energy	84 kJ (20 kcal)
Carbohydrates	4.64 g
- Sugars	2.40 g
- Dietary fibre	1.7 g
Fat	0.17 g
Protein	0.86 g
Vitamin A equiv.	18 μg (2%)
Thiamine (vit. B_1)	0.057 mg (5%)
Riboflavin (vit. B_2)	0.028 mg (2%)
Niacin (vit. B_3)	0.480 mg (3%)
Pantothenic acid (B_5)	0.099 mg (2%)

Vitamin B_6	0.224 mg (17%)
Folate (vit. B_9)	10 μg (3%)
Vitamin C	80.4 mg (97%)
Calcium	10 mg (1%)
Iron	0.34 mg (3%)
Magnesium	10 mg (3%)
Phosphorus	20 mg (3%)
Potassium	175 mg (4%)
Zinc	0.13 mg (1%)

Percentages are relative to US recommendations for adults.
Source: USDA Nutrient Database

25.1. ORIGIN

Capsicum is reported to be native of tropical America (Thomson and Kelly 1957). In India, the introduction of Capsicum is believed to be through the Portuguese in the 17th century.

25.2. CLIMATE

Capsicum is an annual warm season crop. It grows under a temperature of 20°C–25°C. Unfavourable temperature and water supply are the basic reasons for the premature dropping of bud, blossom and fruits. It thrives in areas having a moderate rainfall within the range of 60–120 cm per annum.

25.3. SOIL

Capsicum can be grown in all types of soil. But sandy and loam soil are the best for the cultivation of Capsicum. The soil pH should be within 5.5–6.8.

25.4. LAND PREPARATION

Land should be brought to fine tilth by shallow ploughing 4–5 times with tractor or power tiller giving sufficient intervals between two ploughings. Land should be levelled and divided into small plots and channels to facilitate proper irrigation and intercultural operations. The manures should be applied at the time of land preparation and mixed well with soil by ploughing followed by planking.

25.5. SOWING OF SEED

1. Time of sowing

The seeds should be sown in the nursery bed in the months of September and October.

2. Seed rate

Hybrid variety : 300 g/ha
O.P. variety : 375 g/ha

3. Seed treatment

Seeds should be healthy and free from diseases. They should be treated with any one of the fungicides given in Table 158 to protect the crop from seed–borne or to some extent soil–borne diseases also.

Table 159. Fungicides used for seed treatment of Capsicum

Sr. No.	Names of fungicides/bio-fungicides	Amount (per litre of water)
1.	Carbendazim 50 WP	2.0–3.0 g
2.	M.E.M.C 6 WP	1.0 g
3.	Captan 50 WP	2.0–5.0 g
4.	Mancozeb 75 WP	2.5 g
5.	Trichoderma viridi	4.0 g

4. Raising of seedlings

Seedlings should be raised in nursery beds having measurement of 750 x 120 x 15 cm. The seed bed should be manured at 20 days before sowing of seed at the rate as follows:

FYM or Compost : 15.0–20.0 kg
Nitrogen (N) : 75 g
Phosphorus (P_2O_5) : 75 g
Potassium (K_2O) : 75 g

Seeds should be sown in the nursery bed at a distance of 10 cm between rows and 2 cm between seeds at a depth of 1 cm. The seed bed should be covered with newspaper or straw after watering the seed bed.

Seedlings can also be raised in a wooden box having measurement of 45 x 35 x 75 cm. The wooden box should be filled with leaf mould and sand in the ratio of 1:1. Levelling should be done by giving light pressure on the soil mixture. Seeds should be sown in line of 2 cm in the depth of 0.5–1.0 cm. The soil should be mixed with water in which Carbendazim 50 WP should be mixed @ 2.0 g per litre of water. The box should be covered with gunny bag or newspaper, which should be removed just after germination of the seeds.

25.6. TRANSPLANTING

The seedlings become ready for transplanting in about four weeks when they are 15–20 cm in height and having 4–5 leaves. The seedlings should be transplanted in spacing as follows:

Row to row distance: 60 cm

Plant to plant distance: 45 cm

25.7. MANURING

Capsicum has a long growing season and therefore, requires a judicious use of manures and fertilizers as follows:

Manures

Compost or FYM: 20.0–25.0 t/ha

Fertilizers

Table 160. Recommended amount of fertilizers for Capsicum

Sr. No.	Variety	Amount (kg / ha)		
		Nitrogen (N)	Phosphorus (P_2O_5)	Potassium (K_2O)
1.	Hybrid	200.0	150.0	150.0
2.	O.P	125.0	75.0	50.0

Manuring schedule

a. Hybrid variety

Basal dressing - All quantity of phosphorus and potassium and half quantity of nitrogen.

Top dressing - The remaining amount of nitrogen should be applied in three equal doses at 20, 40 and 55 DAP.

b. O.P. variety

Basal dressing - All quantity of phosphorus and potassium and half quantity of nitrogen.

Top dressing - The remaining amount of nitrogen should be applied in two equal doses at 30 and 45 DAP.

25.8. IRRIGATION

The crop should be irrigated frequently to maintain soil moisture. The maintenance of uniform soil moisture is essential to prevent the pre-mature blossom and fruit drop.

In India, generally 8–9 irrigations are given depending on soil type, humidity and rainfall.

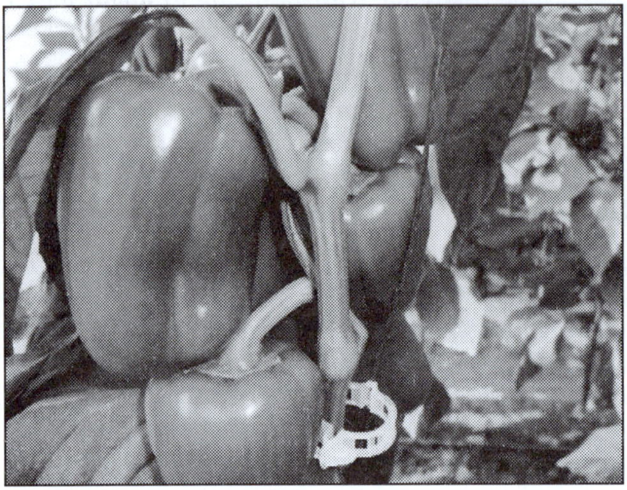

Fig 51. Capsicum

25.9. INTERCULTURAL OPERATION

2–3 shallow hoeing should be given to the soil to kill the weeds and provide soil mulch during early stages of growth. Application of weedicides for controlling the weeds is found effective. Lasso @ 1.5 litre per hectare with one hand weeding or Tok-E 25@ 2 litres per hectare with one hand weeding have been found to be effective in controlling the weeds.

25.10. HARVESTING

The Capsicums should be harvested when they are still green in colour, but, full grown. They should be picked up at frequent intervals, preferably twice a week. The fruits become ready for harvesting within 65–75 DAT.

25.11. YIELD

120–250 quintals per hectare

Capsicum can be stored for about 40 days at 0°C and 95–98 per cent relative humidity (Platenius and Thomson 1939).

25.12. VARIETIES

The important varieties of Capsicum are as follows:

General variety

California wonder, Yellow wonder, King of North, Chinese Giant, Arka Mohini, Arka Gaurav, Arka Basant, Sweet Banana, Midway etc.

Hybrid variety

Bharat, Maha Bharat, Golden Summer, N0-3 (Green Gold), MCPH-11, Nath Hira, MHP-1, MHP-2, MHP-9, MHP-10 etc.

26 CHILLI

Scientific name

Capsicum annum L. var. *longum* (Fruits are long or small, but pungency is high)

Capsicum frutescense (Large and inflated variety with very little pungency)

Family - Solanaceae

Commercial part - Fruit

Indian names - *Hindi* : Lal mirch; *Bengali* : Lanka, Lankamorich; *Gujarati* : Marcha; *Kannada* : Mensina kai; *Malayalam* : Mulaku; *Marathi* : Mirchi; *Oriya* : Lanka; *Punjabi* : Lalmirch; *Tamil* : Milagay; *Telugu* : Mirapa kaya; *Urdu* : Lalmirch.

Chilli is one of the most valuable crops of India. The crop is grown largely for its fruits all over India. It is used in India as a principle ingredient of various curries and chutneys. It is also used for vegetables, spices, condiments, sauces and pickles. Dry chillies are used for curry powder. Pungency in Chillies is due to the active constituent 'Capsaicin', an alkaloid that is extracted from Chillies and is used in medicines. The bright red colour of Chilli at the ripening stage is due to the pigment 'Capsathin'.

26.1. CLIMATE

The Chilli is a plant of tropical and sub–tropical region. It grows well in warm and humid climate and at a temperature of 20°C–25°C. Low moisture in soil during blossom development and fruit formation causes the bud to wither and then the fruit drops. Excessive rainfall is detrimental to the crops because it brings about defoliation and rotting of the plant.

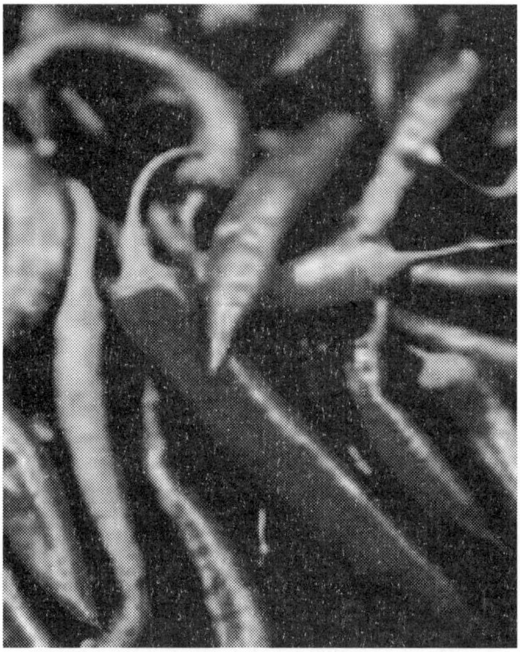

Fig 52. Chilli

26.2. SOIL

Chilli can be grown in all types of soft soils but, the sandy loam, clay loam and loam soils are best suited for it. The soil must be well–drained and well–aerated. Acidic soils are not suitable for Chilli cultivation.

26.3. LAND PREPARATION

The land should be prepared by giving 2–3 ploughings and clod crushing after each ploughing. Compost or FYM @ 10–12 tons per hectare should be spread and mixed well in the soil at least 15–20 days before planting of seedlings. At the last ploughing, Chloropyriphos 1.5 D @ 20.0–25.0 kg per hectare should be applied to the soil to protect the crop from white ants and other soil pests.

26.4. SOWING OF SEED

Time of sowing seed

Chilli can be grown almost at any time of the year and accordingly, seeds can be sown almost all the year round. In south India, the seeds are sown in the month of April–May. Chilli is mainly grown in summer and winter, and the seeds are sown in the nursery bed as follows:

Summer crop : January–February

Winter crop : July–August

Hybrid : August–September, December–January

Seed rate

General variety : 1,000–1,250 g/ha

Hybrid variety : 300–325 g/ha

O.P. variety : 650–750 g/ha

Raising of seedlings

Chilli is raised from the seed. Hussain et al (1970) reported that transplanting gave higher and earlier yield than direct sowing. The seeds should be sown in well–prepared, raised nursery beds. The surface of nursery bed should be well pulverized and heavily manured with decayed leaf mould or FYM. Seeds should be sown at a depth of 0.5 cm and covered with fine soil. Immediately after sowing, nursery bed should be covered with straw or old gunny bag. As soon as the seeds begin to germinate, the covering materials should be removed. After sowing of seeds, watering should be done moderately with a water can. Hand watering should thus be continued every day except on a rainy day. However, care should be taken to avoid over–watering or stagnation of water, as this leads to a fungal disease known as 'damping off', which kills the seedlings. The seedlings should be sprayed regularly with fungicide such as Mancozeb 75 WP @ 2.5 g per litre of water and insecticide such as Acephate 75 WP @ 0.75 g per litre of water, to save the seedlings from insect pest and diseases.

26.5. TRANSPLANTING

The seedlings become ready for transplantation within 6–7 weeks of sowing of seed in the nursery bed. According to Anonymous (1961) and Thomson (Litarture citation), it is ideal to use 6–8 weeks old and 15–20 cm high seedlings for transplanting, after topping them in the nursery bed in the morning of any sunny day at 8–10 days before transplanting. Topping produces better and thicker seedlings, which help to set roots quickly in the field. This operation also keeps off the curl disease caused by Chilli thrips on tender leaves.

The planting distance varies according to the variety, season and soil fertility. The seedlings are generally transplanted in the spacing given below:

Row to row : 45–60 cm

Plant to plant : 45 cm

According to Singh (1953), transplanting of Chilli for the second time, (i) induced earliness in flowering and fruiting, (ii) enhanced the yield to an extent not less than 25 per cent, and (iii) extended the period of longevity. Chillies are also grown with Brinjal (*Solanum melongena*) as a companion crop.

26.6. MANURING

a. General variety

The manuring schedule as per the recommendation of Directorate of Agriculture, Government of West Bengal is as follows:

Manures

Compost or FYM	: 10.0 t/ha
Azophos	: 15.0 kg/ha

Fertilizers

i. Macronutrients

Nitrogen (N)	: 90.0 kg/ha
Phosphorus (P_2O_5)	: 60.0 kg/ha
Potassium (K_2O)	: 60.0 kg/ha

ii. Micronutrients

Zinc	
Zinc Sulphate	: 25.0 kg/ha
Boron	
Borax	: 10.0 kg/ha
Molybdenum	
Ammonium molybdate	: 0.5 kg/ha

Method of application

Basal dressing - All quantity of manures, azophos and micronutrient, total amount of phosphorus, half amount of nitrogen and potassium

Top dressing - One fourth quantity of nitrogen and potassium at 21 DAP and remaining amount of nitrogen and potassium at 40 DAP

b. Hybrid variety

Manures: Compost or FYM: 25.0 t/ha

Fertilizer

(i) Macronutrients

Nitrogen (N)	: 135.0 kg/ha
Phosphorus (P_2O_5)	: 50.0 kg/ha

Potassium (K_2O) : 75.0 kg/ha

(ii) Micronutrients

Zinc

Zinc Sulphate : 25.0 kg/ha

Boron

Borax : 10.0 kg/ha

Molybdenum

Ammonium molybdate : 0.5 kg/ha

Method of application

Basal dressing: All quantity of manures, azophos and micronutrient, total amount of phosphorus, 45.0 kg of nitrogen and 35.0 kg of potassium.

Top dressing

- At 45 DAP: 18.0 kg of nitrogen and 8.0 kg of potassium
- At 60 DAP: 18.0 kg of nitrogen and 8.0 kg of potassium
- At 75 DAP: 18.0 kg of nitrogen and 8.0 kg of potassium
- At 90 DAP: 18.0 kg of nitrogen and 8.0 kg of potassium
- At 105 DAP: 18.0 kg of nitrogen and 8.0 kg of potassium

26.7. IRRIGATION

Chilli is grown both as rainfed and irrigated crop. First irrigation is given after the transplanting and subsequent irrigations are given at intervals of 5–7 days depending on the weather and conditions of soil during summer and rainy season, and after every 10–15 days in winter. The maintenance of uniform soil moisture is essential to prevent blossom and fruit drops. Flower drop in Chillies is a great problem and it depends on the high temperature, low moisture availability, shading and light intensity. Flowers and/or flower bud abscission was increased under short day (day length 12 hours) and high temperature (28°C–33°C). Spraying the Chilli with Planofix (NAA) at 10 ppm at flower initiation stage and 15 days later (peak flowering) reduces the flower drop and increases the fruit set.

26.8. INTERCULTURAL OPERATION

2–3 shallow hoeings should be given to the soil to kill the weeds and provide soil mulch during early stages of the growth. Application of weedicides for controlling the weeds is found effective. Alachlor 50 EC (Lasso Alachlor, Dorado etc) @ 1.5 litre per hectare with one hand weeding or Tok-E 25@ 2 litres per hectare with one hand weeding have been found to be effective in controlling the weeds.

26.9. HARVESTING

Chilli is harvested in about 60–70 days in North India and 90–105 days in South India. Chillies are harvested at different stages on the basis of uses as follows:

- **Vegetable purposes**: Chillies are generally harvested while they are still green, but full grown.
- **Canning purposes**: Chillies are harvested at red stage for canning purpose. Chillies used for drying are harvested at full–ripe stage.

26.10. YIELD

The yield varies according to the system of cultivation. The yield of dry Chillies of rainfed crop is 500–1,000 kg and that of irrigated crop is 1,500–2,500 kg per hectare. The proportion of dry to fresh ripe Chillies varies from 25–40 per cent.

26.11. VARIETIES

- **Improved varieties**

N.P. 46, Pnsa Jwaja, G-3, CA-960, Pant C-t X-235, AKC-79-18, Parbhani Tejas

- **Local varieties :** Achalpuri, Bhavapuri, Vashirn, Malkapuri

26.12. PLANT PROTECTION

Insect pests

1. Chilli thrips

Scientific name - *Scirto-thrips dorsalis* H

Thrips is the common insect pest, which affects the crop throughout its life cycle. But they are more severe when plants begin to flower.

Nature of damage

The thrips, small insect, suck the sap from the foliage and lacerate the leaf tissue, which result in curling of leaves and fall down of flowers prematurely.

Control measures

- Hand picking and destruction of eggs and larva is the best method of controlling the thrips.
- The insect–affected parts of the plant should be clipped and destroyed by burning or by burrowing in the soil.
- Thrips can effectively be controlled by spraying the crop with any one of the insecticides given in Table 161.

Table 161. Insecticides used for controlling the Chilli thrips

Sr. No.	Name of insecticide	Amount (per litre of water)
1.	Acephate 75 WP	0.75 g
2.	Cartap hydrochloride 50 SP	1.0 g
3.	Malathion 50 EC	2.0 ml

2. Pod borer

Scientific name - *Spodoptera (Prodemia) litoralis, Spodoptera exigua, Heliothis armigera*

Nature of damage

The caterpillar eats leaves and later on bores the pod, which results in the deterioration of quality and market price of the product.

Control measures

- The insect–affected Chillies should be removed and destroyed by burning or by burrowing under the soil.
- Hand picking and destruction of eggs and larva is the best method of controlling the thrips.
- The chemical control measures are the timely spraying on the crop with any one of the insecticides given in Table 162, starting from flower bud initiation.

Table 162. Insecticides used for controlling the Pod borer of Chilli

Sr. No.	Name of insecticide	Amount (per litre of water)
1.	Acephate 75 WP	0.75 g
2.	Thiomethoxam 25 WG	0.2 g
3.	Spinosad 45 SC	0. 25 ml
4	Bifenthin 10 EC	1.5 ml

3. Aphids

Scientific name - *Aphis gossypi*

Aphid is most destructive insect causing lower yield.

Nature of damage

Aphid sucks the sap from the plants. They generally attack the crop in winter months and at the later stages of the crop. The quality of the produce is spoiled by imparting blackish colour to the calyx and pods. They also serve as a Vector to Virus (Shanmugavelu 1990).

Control measures

- The aphid–infested plants should be destroyed by burning or by burrowing under the soil as soon as they are noticed.
- Use of yellow trap, Feroman trap etc. are found effective in controlling the aphid.
- Judicial application of nitrogenous fertilizer helps in checking the insect population.
- Early sowing of crop and sowing the crops such as Maize (*Zea mays*), Mustard (*Brassica* sp) etc. as border crop helps in checking the incidence of the Aphid.
- Aphid can effectively be controlled by spraying the crop with any one of the insecticides given in Table 163 at an interval of 5 days.

Table 163. Insecticides used for controlling the Aphid

Sr. No.	Name of insecticide	Amount (per litre of water)
1.	Dimethoate 30 EC	2.0 ml
2.	Acephate 75 WP	0.75 g
3.	Cartap hydrochloride 50 SP	1.0 g
4.	Clothianidin 50 WDG	0.1 g

- Soil application of the insecticides given in Table 164, in two doses i.e., half at the time of planting and other half at the time of earthling up is also effective against aphid.

Table 164. Insecticides used as soil application for controlling the Aphides of Chilli

Sr. No.	Name of insecticide	Amount (kg per hectare)
1.	Phorate 10 G	12.5
2.	Carofuran 3 G	17.5

4. Mites

Scientific name - *Hemitarsonamus latus*

Nature of damage

Mites attack the Chilli shoot and inflorescence causing malformation and shedding of flower.

Control measures

- The crop belonging to the family Solanaceae should not be cultivated in succession. Proper crop rotation should be followed.
- Maize (*Zea mays*) and Jowar (*Sorghum vulgare*) should be cultivated as trap crop near the Chilli field and weeds should be controlled.
- Spraying of cold water on the plant is found effective in controlling the mites.
- Application of Neem cake @ 150.0–187.5 kg per hectare reduces the attack of mite.
- The mites can effectively be controlled by spraying the crop with any one of the insecticides given in Table 165 plus Planofix to check the premature drop of flower due to insecticides:

Table 165. Insecticides used for controlling the Mites of Chilli

Sr. No.	Name of insecticide	Amount (per litre of water)
1.	Sulphur 80 WP	2.5 g
2.	Neem oil 1,500 ppm	3.0 ml
3.	Dicofol 18.5 EC	1.0 ml

Diseases

1. Damping off

C.O.: *Phytopthora* sp, *Phythium* sp

Damping off is a serious disease of Chilli seedlings and mainly occurs in nursery bed. Pre–emergence and post–emergence damping off occurs in the nursery bed.

Symptoms

The disease–infected seedlings rot at ground level and then the plants fall over ground. The seedlings die in patches.

Control measures

- The seed bed should be treated with Formalin before sowing the seeds and keeping the seed bed covered with polythene sheet for a week before sowing the seed.
- The seeds should be treated with hot water (30 min at 52°C) or Cerasan or Agrosan G.N. or Carbendazim 75 WP or *Trichoderma viridi* before sowing them.
- The seedlings in the nursery should be sprayed with any one of the fungicides given in Table 166 at a regular interval.

Table 166. Fungicides used for controlling the Damping off of Chilli

Sr. No.	Names of fungicides	Amount (per litre of water)
1.	Carbendazim 50 WP	1.0 g
2.	Copper oxychloride 50 WP	5.0 g
3.	Bacillus thurengiensis (BT)	1.0 g

2. Bacterial leaf spot or Bacterial blight

C.O.: *Xanthomonas vesicatoria*

Bacterial blight is a bacterial disease of Chilli.

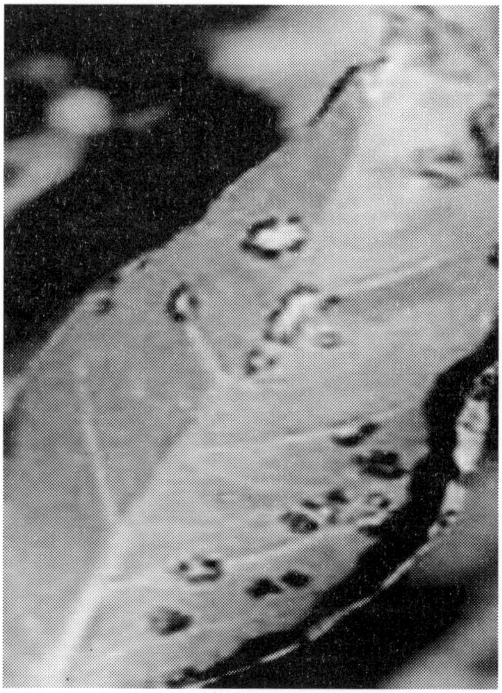

Fig. 53. Bacterial leaf spot or Bacterial blight

Symptoms

Small dark, greasy spots are formed on the leaf, petiole and tender parts of the plant. Water–soaked spots appear on green fruits. The spots coalesce to form irregular lesions. Severely–affected leaves become chlorotic and fall off. Petioles and stems are also affected. In severe cases the leaf may drop off and can cause considerable loss to the crop.

Control measures

- Cultivation of disease–resistant varieties such as Arka Gaurav, Arka Basant etc. is found to be effective in controlling the incidence of this disease.
- Spraying Agrimycin - 100 @ 200 ppm plus Copper oxychloride 50 WP @ 5.0 g per litre of water or Steptocycline @ 1.0 g per 10 l of water, controls the disease effectively (Sreeramulu, 1976).

3. Anthracnose

C.O.: *Collectotrichum capsici*

Anthracnose disease is one of the major economic constraints to Chilli production worldwide, especially in tropical and subtropical regions.

Symptom

Dark sunken spots are formed on fruits and pink or dark coloured dots appear in the centre of the sunken spots. Due to these spots, the fruits rot and fall. The fungus may cause 'die back' of the twigs also (Mehta and Verma 1968). Die back disease mainly attacks the upper portion of the plants and then gradually spreads downwards; as a result of which the branches dry up. Moist weather, shade and heavy dew favour the occurrence of the disease (Chauhan1967).

Control measures

- Crop rotation with non–solanaceous crop should be practiced.
- Removing and burning of attacked plants or branches is a good method for controlling the anthracnose disease.
- The seed should be free from disease. For this, seeds should be collected from healthy and disease–free crop.
- The seeds should be treated with any one of the seed treatment chemicals given in Table 167.

Table 167. Seed treatment chemicals used for controlling the Anthracnose of Chilli

Sr. No.	Name of Seed treatment chemical	Amount (per litre of water)
1.	Carbendazim 50 WP	2.0–3.0 g
2.	M.E.M.C 6 WP	1.0 g
3.	Captan 50 WP	2.0–5.0 g
4.	Mancozeb 75 WP	2.5 g
5.	Trichoderma viridi	2.1 g

- Spraying the crop with any one of the fungicides given in Table 168 is found to be effective in controlling the anthracnose disease:

Table 168. Fungicides Used for controlling the Anthracnose of Chilli

Sr. No.	Names of fungicide	Amount (per litre of water)
1.	Carbendazim 50 WP	1.0 g
2.	Copper oxychloride 50 WP	5.0 g
3.	Mancozeb 75 WP	1.5 g

4. Leaf curl

C.O.: Virus

Leaf curl is a disease caused by virus.

Symptoms

The disease–affected leaves become small in size accompanied by downward curling. The leaves may fall off in case of severe attack. The disease usually spreads through insect vectors such as thrips and aphids etc.

Control measures

The disease–affected plants should be pulled and destroyed by burning or burrowing under the soil. The disease–resistant variety (e.g., Pusa Red, Pusa Jawala, Suryamukhi) should be cultivated. Good drainage facility helps in reducing the incidence of the leaf curl disease.

Control of insect vector, by spraying the crop with any one of the insecticides given in Table 169, helps to check the spread of the disease.

Table 169. Insecticides used for controlling the insect vector of leaf curl

Sr. No.	Name of insecticide	Amount (per litre of water)
1.	Acephate 75 WP	0.75 g
2.	Cartap hydrochloride 50 SP	1.0 g
3.	Malathion 50 EC	2.0 ml

5. Dieback

C.O.: *Collectotrichum capsici*

Die back disease is observed at flowering stage of Chilli crop.

Symptoms

Drying of branches starts from the tip and proceeds to the base; hence, called as dieback.

Control measures

- The seed should be treated with seed treatment chemical before sowing.
- The disease–affected parts of the plant should be removed and destroyed by burning or by burrowing under the soil.
- Spraying the crop with any one of the fungicides given in Table 170 is found to be effective in controlling the dieback disease.

Table 170. Fungicides used for controlling the Die back of Chilli

Sr. No.	Name of fungicide	Amount (per litre of water)
1.	Carbendazim 50 WP	1.0 g
2.	Copper oxychloride 50 WP	5.0 g
3.	Mancozeb 75 WP	2.5 g

6. Chilli mosaic

C.O.: Chilli mosaic virus

Symptoms

The affected plant develops mottling of yellow and green colour on the leaves. The margins may be curled instead of being flat and smooth. The affected leaves develop depression, raised areas and puckering on the leaf blade. The disease–affected plants get stunted and produce few flowers. The fruits are considerably reduced in size, shrivelled and deformed.

Control measures

- The nursery bed and field should be kept free from weeds by regular weeding.
- The planting of Chillies close to cucumber, tobacco, tomato etc. should be avoided.

- Infected plants should be roughed out and destroyed by burning or by burrowing under the soil.
- The spread of the disease can be reduced by controlling the vector by spraying any one of the insecticides given in Table 171.

Table 171. Insecticides used for controlling the insect-vector of Chilli mosaic

Sr. No.	Name of insecticide	Amount (per litre of water)
1.	Acephate 75 WP	0.75 g
2.	Cartap hydrochloride 50 SP	1.0 g
3.	Malathion 50 EC	2.0 ml

Spraying should be stopped at least 20 days before picking of fruits.

7. Powdery mildew

C.O.: *Leveillula taurica*

The powdery mildew disease is common in all Chilli–growing areas and occurs particularly during the winter season. Cloudy weather is highly favourable for the disease development.

Symptoms

The symptoms of powdery mildew disease first appear on leaves and then on other green parts of the plant. In initial stages, small, white circular powdery spots appear on the upper surface of leaves, which gradually cover the entire leaf, stem, petiole and fruits. In advanced stages of the disease, the leaf appears as if covered by talcum like powder. The plant assumes a greyish appearance and then turns yellow and fruits are shed profusely. The disease–affected plants lose their vigour and remain barren, and the yield is very much reduced.

Control measures

- The crop residues should be destroyed by burning as the fungus perpetuates in the debris of the plant.
- Late sowing should be avoided.
- The disease can be controlled by dusting the crop with sulphur 80 WP @ 3.0 kg per hectare or spraying the crop with any one of the fungicides given in Table 172, starting just on the appearance of the disease, at fortnight interval.

Table 172. Fungicides used fo controlling - Powdery mildew of Chilli

Sr. No.	Name of fungicide	Amount (per litre of water)
1.	Tridemorph 80 EC	0.5 ml
2.	Sulphur 80 WP	2.0–3.0 g
3.	Hexaconazol 5 EC	1.5 ml
4.	Myclobutanil 10 WP	0.5 g

8. Fruit rot

C.O.: *Phytopthora capsici*

Symptoms

The disease affects ripe fruits, which turn red, become watery and begin to rot. A small, black, circular spot appears on the skin of the fruit and spreads in the direction of the long axis of fruit, thus becoming more or less elliptical. Numerous spots are formed on the discoloured area. In advanced stages, the seeds are covered by fungal mat with black dots. Green fruits are generally not attacked. Affected fruits become deformed, white in colour and lose their pungency.

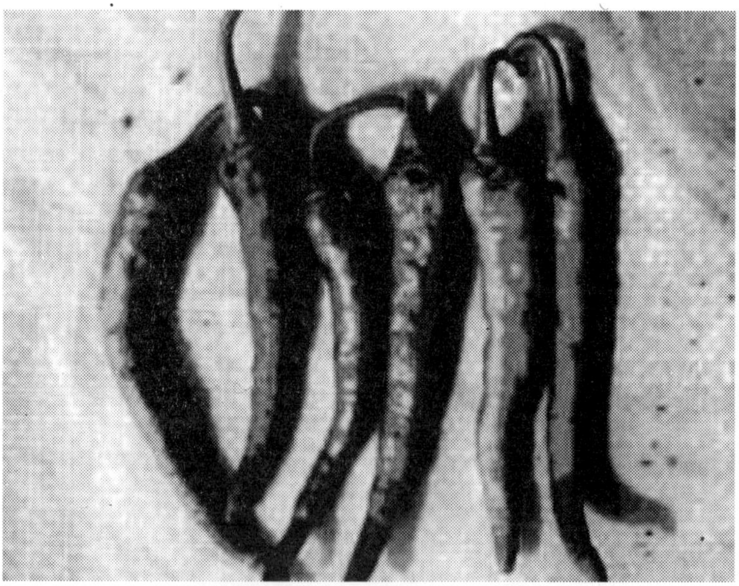

Fig 54. Fruit rot of Chilli

Control measures

- The continuous cultivation of Chili in the same field should be avoided.

- Seeds should be collected from apparently healthy pods for sowing and should be treated with seed treatment chemicals before sowing.
- The growing crop like turmeric, brinjal, cotton, bengal gram, jute and tomato near by the chilli field should be avoided.
- The disease can be controlled by spraying any one of the fungicides given in Table 173 in the nursery to avoid transmission of disease from nursery to main field and in crops in the main field at 115, 130 and 140 days after transplanting.

Table 173. Fungicides used for controlling the fruit rot of Chilli

Sr. No.	Name of fungicide	Amount (per litre of water)
1.	Carbendazim 50 WP	1.0 g
2.	Clorothalonil 75 WP	2.0 g
3.	Mancozeb 75 WP	2.5 g
5.	Cptan 50 WP	2.5 g

27 BASIL

Scientific name - *Ocimum basilicum, Ocimum sanctum*

Family - Lamiaceae

Commercial part - Leaf

Indian names - *Bengali* : Babui tulsi or Kalotulsi; *English* : Basil; *Hindi* : Barbar or Tulsi; *Assamese* : Tulasii; *Gujarati* : Sabje; *Kannada* : Kama kasturi, Ramkasturi, Tulasigidda, Tulasiya sasyajati or Vishnu tulasi; *Malayalam* : Pachcha or Sivatulasi; *Marathi* : Sabja or Tulas; *Oriya* : Dhalatulasi; *Punjabi* : Tulsi; *Tamil* : Tirunirrippachai, Tiviragandam or Tulasi; *Telugu* : Oddhi, Rudrajada or Tulsi-chettu;

Fig. 55. Basil

The word 'Basil' comes from the Greek *basileus*, meaning 'king'.

Basil originated in India, where it was regarded as a sacred herb. Basil is an attractive annual, about 18 inches tall with fairly broad leaves that are light–

green in colour. The flowers are small, white in colour, and appear in spikes. There are several species of cultivated Basil, one having purple leaves. Two species of Tulsi are cultivated on large scale in India as follows:

1. Indian basil (*Ocimum basilicum*) - Indian basil is cultivated for oil and 250 ton oil is produced annually in North India and the oil is consumed in flavour and fragrance industries in India and abroad. Indian basil is a short duration crop and can be cultivated in wide range of soil and climatic conditions as a short duration crop (75–80 days) with an average yield of 40–50 kg oil/acre.

2. Sacred basil / Pusa basil (*Ocimum sanctum*) - Pusa basil is mainly cultivated for dry leaves, which is a constituent of herbal medicine and teas.

27.1. CLIMATE

Basil thrives in warm and sunny climate. Basil needs plenty of sunlight to grow and flower. It is susceptible to frost and cold. It is highly subjected to heat and cold stress.

27.2. SOIL

Basil grows in different kinds of soil. But, it prefers moist, rich and well-drained soil with a pH of 6.0.

27.3. LAND PREPARATION

Land should be brought to fine tilth by shallow ploughing 4–5 times with tractor or power tiller giving sufficient intervals between two ploughings. Land should be levelled and divided into small plots and channels to facilitate proper irrigation and intercultural operations. The manures should be applied at the time of land preparation and mixed well with soil by ploughing followed by planking.

27.4. PLANTING

Basil may be grown in the field from seeds. Seeds should be sown in well-prepared nursery. The seed rate should be 6.75 kg/ha to produce seedlings for transplanting. Plants emerge in 8–14 days after sowing. Basil may also be grown from transplants started in the greenhouse in late March to early April for the piedmont and coastal countries and mid–April to early May for the mountains. Basil also roots readily from cuttings. The seedlings should be transplanted in the main field at spacing of 2–3 ft (60-90 cm) between rows and 6–12 inches (15-30 cm) within the row. The seedlings should be trimmed to encourage branching.

27.5. MANURING

Manures: Compost or FYM: 10.0 t/ha

Fertilizers

Nitrogen (N)	: 112.5 kg/ha
Phosphorus (P_2O_5)	: 112.5 kg/ha
Potassium (K_2O)	: 125.0 kg/ha

The manures and fertilizers should be applied by broadcasting method at the time of land preparation and mixed well by cross ploughing followed by planking.

27.6. IRRIGATION

Basil cannot tolerate moisture stress. Irrigation should be provided as per the requirement of the plant. The plant should be provided regular water supply through drip or overhead irrigation.

27.7. INTERCULTURAL OPERATION

The land should be kept free from weeds. Weeds pose problem in the growth and development of crop. Regular weeding should be undertaken to keep the field free from weeds.

27.8. CULTIVARS /VARIETIES

Genovese, Spicy Glove, Holy basil, African basil, Dark opal purple, Sian queen, Serata basil, Purple Ruffles, Pesto perpetuo etc.

27.9. HARVESTING

The harvesting of leaves can be started when Basil reaches 6–8 inches in height. Harvesting of Basil is important not only to promote the growth of new stems but to ensure that the plant does not flower too early. Once the herb starts to flower, it stops producing new leaves and the existing leaves may become slightly bitter in taste. The leaves can be used fresh, dried, or preserved in oil (must be refrigerated) or vinegar.

28 MARJORAM

Scientific name - *Marjorana hortensis*

Family - Lamiaceae

Commercial part - Leaf and Flower top

Indian names - *Hindi*: Marwa; *Bengali* : Murru, Deccan, Muruva; *Kannada* : Maruga, Kumaon bantulsi; *Malayalam* : Maruva; *Punjabi* : Marwa; *Sanskrit* : Maru, Sinshi, Murwo; *Tamil* : Maruvu; *Urdu* : Marva khusha.

Marjoram, also called Sweet marjoram and Knotted marjoram, is a tender perennial grown as an annual.

28.1. SOIL

Marjoram grows well in different types of soil. Rich, well–drained soil with pH between 6.7 and 7.0 is ideal for the cultivation of Marjoram. Marjoram can thrive in poor soil that is well–drained.

28.2. LAND PREPARATION

Land should be brought to fine tilth by shallow ploughing 4–5 times with tractor or power tiller giving sufficient intervals between two ploughings. Land should be levelled and divided into small plots and channels to facilitate proper irrigation and intercultural operations. The manures should be applied at the time of land preparation and mixed well with soil by ploughing followed by planking.

28.3. PROPAGATION

Marjoram is propagated either from seed in spring or from cuttings in the summer or root divisions in the fall season. The sowing time of Marjoram is as follows:

1. Wild or pot marjoram - The wild marjoram can be sown directly in seed bed in the month of April. The seeds can be sown 2 cm deep and 15 cm apart. The seedlings can be thinned to 30 cm apart when the seedlings emerge two weeks later.

2. Sweet/ knotted marjoram - The seeds of Sweet marjoram can be sown indoors in pots during the month of March and gradually hardening off the plants in April after which they can be potted up or planted outside 6 inches apart. Space rows 45-60 cm apart.

The seedlings should be planted 45–60 cm apart in rows and at a distance of 15 cm from plant to plant.

Fig 56. Marjoram

28.4. MANURING

Manures: Compost or FYM: 10.0 t/ha

Fertilizers

Nitrogen (N)	: 112.5 kg/ha
Phosphorus (P_2O_5)	: 112.5 kg/ha
Potassium (K_2O)	: 125.0 kg/ha

The manures and fertilizers should be applied by broadcasting method at the time of land preparation and mixed well by cross ploughing followed by planking.

28.5. INTERCULTURAL OPERATION

The land should be kept free from weeds. Weeds pose problem in the growth and development of the crop. Regular weeding should be undertaken to keep the field free from weeds.

28.6. IRRIGATION

Marjoram cannot tolerate water stress. Irrigation should be provided as per the requirement of the plant. The plant should be provided regular water supply through drip irrigation.

28.7. HARVESTING

Marjoram is ready for harvest 60 days after sowing. The individual Marjoram leaves are harvested fresh throughout the growing season. The largest amounts are harvested by cutting back the entire Marjoram plant just before it flowers. The fresh leaves of Marjoram should be kept in plastic bag and stored in the refrigerator for 3–4 days. Dry leaves and flowers should be stored in a warm place and in an airtight container.

28.8. VARIETIES

Aureum, Aureum crispum, Compactum, Thumbles variety, Kaliteri, White anniversary, Italian oregano (*O. x majoricum*) etc.

29 OREGANO

Scientific name - *Origanum vulgare*

Commercial part - Leaf and Flowering top

Family - Lamiaceae

Indian names- *Hindi* : Sathra, Mirzanjosh; *Malayalam* : Kattumaruva; *Punjabi* : Mirzanjosha; *Urdu* : Mirzanjosha.

Oregano is a common species of *Origanum*, a genus of the Mint family (Lamiaceae). Oregano is a perennial herb, 20–80 cm tall, with 1–4 cm long opposite leaves. It is native to warm, temperate western and south–western Eurasia and the Mediterranean region. Although grown as an annual in colder climates, it often does not survive the winter months. The ancient Greeks considered Oregano to be the 'joy of the mountain'. Oregano herb is widely used as a spice. It is principally employed in the cuisines as a seasoning to enhance the flavour.

Fig 57. Oregano

Table 174. Nutritive value of Oregano

Principle	Nutrient value
Energy	306 kcal
Carbohydrates	64.43 g
Protein	11 g
Total Fat	10.25 g
Cholesterol	0 mg
Dietary Fibre	42.8 g
Vitamins	
Folates	274 μg
Niacin	6.22 mg
Pantothenic acid	0.921 mg
Pyridoxine	1.210 mg
Riboflavin	0.320 mg
Thiamin	0.341 mg
Vitamin A	6,903 IU
Vitamin C	50 mg
Vitamin E	18.86 mg
Vitamin K	621.7 μg
Electrolytes	
Sodium	15 mg
Potassium	1,669 mg
Minerals	
Calcium	1576 mg
Copper	0.943 mg
Iron	44 mg
Magnesium	270 mg
Manganese	4.667 mg
Zinc	4.43 mg
Phytonutrients	
Carotene-ß	4112 μg
Crypto-xanthin-ß	11 μg
Lutein-zeaxanthin	862 μg

Source: USDA National Nutrient data base

Oregano thrives in well–drained, sandy, and relatively dry soils. Oregano grows in a pH range between 6.0 (mildly acid) and 9.0 (strongly alkaline) with a preferred range between 6.0 and 8.0.

Oregano can be propagated from seeds, stem cuttings, or root division. Oregano seeds require some light to germinate, so the seeds should be covered slightly with soil.

The young seedlings can be planted during the early summer or in the spring. Plants must be spaced at least 30 cm apart from each other for optimal growth. Oregano grows very slowly and the soil must be weeded on a regular basis to ensure maximum growth of the crop during the growing season.

The leaves of Oregano are used for flavouring foods. They retain their flavour better in hot dishes, if added towards the end of cooking. Dried Oregano has a stronger taste than fresh. In order to keep the fragrance and aromatic flavour intact, Oregano leaves are generally used just before preparing the recipe.

30 MINT

Scientific name - *Mentha arvensis* Linn.

Family - Lamiaceae

Commercial part - Leaf

Indian names - *Bengali* : Pudina; *Hindi* : Pudina; *Gujarati* : Pudina; *Marathi* : Pudina; *Punjabi* : Pudina; *Urdu* : Pudina; *Kashmiri* : Pudyanu; *Malayalam* : Muthina; *Tamil* : Pudina; *Telugu* : Pudina.

Mentha (also known as Mint, from Greek *mínth* Linear B *mi-ta*) is a genus of flowering plants in the family Lamiaceae (Mint family). Mint is aromatic, almost exclusively perennial, rarely annual herb. India is the largest grower of pudina and producer of menthol and Mint products in the world accounting for 85% of the global production. 80% of the crop in India is grown in Uttar Pradesh. The remaining 20% is grown in Punjab and Haryana.

Table 175. Nutritive values of Mint (Pudina)

Parameters	Value per 100 g
Moisture	84.900 gm
Protein	4.800 gm
Fat	0.600 gm
Minerals	1.900 gm
Fibre	2.000 gm
Carbohydrates	5.800 gm
Energy	48.000 K cal
Calcium	200.000 mg
Phosphorus	62.000 mg
Iron	15.600 mg

Minerals and trace elements	
Magnesium	60.000 mg
Copper	0.180 mg
Manganese	0.570 mg
Zinc	0.440 mg
Chromium	0.008 mg
Sulphur	84.000 mg
Chlorine	34.000 mg
Oxalic Acid	33.000 mg
PhytinPhosphorus	4.000 mg
Carotene	1620.000 μg
Thiamine	0.050 mg
Riboflavin	0.260 mg
Niacin	1.000 mg
Folic Acid (Free)	9.700 μg
Folic Acid (Total)	114.000 μg
Vitamin C	27.000 mg

30.1. CLIMATE

Mint grows well in temperate climatic conditions. Tropical climate is not suitable for the cultivation of Mint. It grows best in wet environment and moist soil, and prefers partial shade.

30.2. SOIL

Mint can grow in different types of soil. It grows well in loam to sandy loam, well–drained, well–aerated and loose–textured soil. Clay soils are not suitable.

30.3. LAND PREPARATION

Land should be brought to fine tilth by shallow ploughing 4–5 times with tractor or power tiller giving sufficient intervals between two ploughings. Land should be levelled and divided into small plots and channels to facilitate proper irrigation and intercultural operations. The manures should be applied at the time of land preparation and mixed well with soil by ploughing followed by planking.

30.4. PROPAGATION

Most Mints can be easily propagated by planting cuttings from the runners of healthy Mints. Mint is planted in the months of January to February in Tarai areas or before the start of the rainy season.

30.5. MANURING

Manures: Compost or FYM: 10.0 t/ha

Fertilizers

Basal dressing

Nitrogen (N)	: 50.0 kg/ha
Phosphorus (P_2O_5)	: 75.0 kg/ha
Potassium (K_2O)	: 37.0 kg/ha

Top dressing

Nitrogen (N): 75.0 kg /ha

Fig 58. Mint

30.6. INTERCULTURAL OPERATION

Mentha needs frequent interculturing weeding and hoeing in early stages of its growth. Afterwards, it needs no weeding as Pudina plants spread very fast and cover the soil.

30.7. IRRIGATION

Mentha crop requires considerable moisture, well–distributed throughout the entire growing season. Light and frequent irrigations are recommended. The crop should be irrigated weekly during summer season.

30.8. HARVESTING

Harvesting is normally done 2–3 times in a year (120 days) as follows:

- First harvesting should be done on the onset of monsoon (May–June).
- Second harvesting should be done well after monsoon is over (September–October).
- Third harvesting should be done in the month of November–December.

30.9. USES

Mint is used for flavouring meat, fish, sauces, soups, stews, vinegar, tea, tobacco and cordials. Mint oil is used for the production of natural menthol. Dementholised oil is used for flavouring mouth washes, tooth paste and pharmaceutical preparations.

30.10. DISEASES

Table 176. Diseases of Mint

Name of disease	Causal organism(C.O.)
Fungal diseases	
Anthracnose	*Sphaceloma menthae*
Black stem rot	*Phoma strasseri*
Leaf blight	*Cephalosporium* sp
Phoma leaf spot	*Phoma exigua*
Powdery mildew	*Erysiphe* spp. *Erysiphe cichoracearum*
Ramularia leaf spot	*Ramularia menthicola*
Rust	*Puccinia menthae, Puccinia angustata*
Septoria leaf spot	*Septoria menthae*
Stem and stolon canker	*Rhizoctonia solani, Thanatephorus cucumeris* [teleomorph]
Stolon decay	*Fusarium solani, Nectria haematococca* [teleomorph]
Verticillium wilt	*Verticillium albo-atrum, Verticillium albo-atrum* var. *Menthae, Verticillium dahliae*
White mould stem rot	*Sclerotinia sclerotiorum*
Viral diseases	
Spotted wilt	Tomato spotted wilt virus

Control measures

- The fungal disease can be controlled by spraying the fungicides given in Table 177.

Table 177. Fungicides used for controlling the fungal diseases of mint

Sr. No	Name of fungicide	Amount (per litre of water)
1.	Carbendazim 50 WP	1.0 g
2.	Clorothalonil 75 WP	2.0 g
3.	Mancozeb 75 WP	2.5 g
5.	Captan 50 WP	2.5 g
6.	Copper oxychloride 50 WP	5.0 g
7.	Metalaxyl 35 WS	0.5 g

- The viral disease can be controlled by controlling the insect vector with the application of any one of the insecticides given in Table 178.

Table 178. Insecticides used for controlling the insect vector of viral disease

Sr. No.	Name of insecticide	Amount (per litre of water)
1.	Acephate 75 WP	0.75 g
2.	Cartap hydrochloride 50 SP	1.0 g
3.	Malathion 50 EC	2.0 ml

31 PARSLEY

Scientific name - *Petroselinum crispum*

Family - Apiaceae

Commercial part - Seed

Indian names - *Hindi* : Ajmood; *Kannada* : Achu-Mooda; *Malayalam* : Seema malli.

Parsley or Garden parsley (*Petroselinum crispum*) is a species of *Petroselinum* in the family Apiaceae, native to the central Mediterranean region (southern Italy, Algeria and Tunisia), naturalized elsewhere in Europe, and widely cultivated as an herb, a spice and a vegetable. The plant is a biennial herb with aromatic, crinkle bright green leaves, small greenish–yellow coloured flowers and produces small brown seeds. Parsley is a native of the Mediterranean region, but it quickly spread throughout the world as a popular culinary essential.

Fig 59. Parsley

31.1. CLIMATE

Parsley does best in a sunny area, which receives direct light for 6–8 hours a day, although it can tolerate some light shade.

31.2. SOIL

Parsley grows in different types of soil. But it grows well in well–drained soil that is fairly rich in organic matter having a pH range of 6.0–7.0.

31.3. PROPAGATION

Parsley is propagated through seeds. A day before planting, the seeds should be soaked overnight in water to help the germination process. The seeds should be sown in a nursery bed measuring 3 m length, 0.8 m width and 15 cm height. The nursery bed should be prepared by mixing forest soil, sand and compost in 1:1:1 ratio. The seeds should be sown in rows 5 cm apart. After sowing, the seeds should be covered with fine compost. The nursery beds should be irrigated three times a day with a rosecan.

Seed rate : 62.5 g/ha

The seedlings can be transplanted in the main field 45 days after sowing at 30 cm distance between rows and between plants.

31.4. MANURING

Manures: Compost or FYM: 5.0 t/ha

Fertilizers

Nitrogen (N)	: 135.0 kg/ha
Phophorous (P_2O_5)	: 135.0 kg/ha
Potassium (K_2O)	: 135.0 kg/ha

Manures should be applied at the time of land preparation. The total amount of phosphorus and potassium and half amount of nitrogen should be applied as basal dressing at the time of last ploughing and mixed well in the soil by cross ploughing followed by planking. The rest amount of nitrogen should be applied as top dressing at 45 DAP.

31.5. INTERCULTURAL OPERATION

Parsley needs frequent intercultural operations such as weeding and hoeing in early stages of its growth. Afterwards, it needs no weeding as Parsley spreads very fast and covers the soil.

31.6. IRRIGATION

Parsley crop requires considerable moisture, well–distributed throughout the entire growing season. Light and frequent irrigations are recommended. The crop should be irrigated weekly during summer season.

31.7. HARVESTING

Parsley generally matures 70–90 days after the seeds germinate and the stems of the plant may be harvested throughout the season. Harvesting is done by cutting the leaves from the outer portions of the plant.

31.8. VARIETIES

Italian flat-leaf parsley, Hamburg parsley, Japanese parsley etc.

32 AJOWAN

Scientific name -*Trachyspermum ammi* [L] Sprague

Syn - *Carum copticum* Heirn

Family - Apiaceae

English name - Bishop's Weed

Commercial part - Seed

Indian names - *Bengali* : Jowan or Joan; *Hindi* : Ajowan; *Gujarati* : Yavan; *Kashmiri* : Jawind; *Kannada* : Oma; *Malayalum* : Omum; *Marathi* : Onva; *Oriya* : Juani; *Urdu* : Ajowain; *Sanskrit* : Ajamoda Yavanika; *Tamil* : Omum; *Telugu* : Vamu.

Fig 60. Ajowan - 1. Plant 2. Seed

Ajowan seeds are an Indian spice. Ajowan seeds, like other spices, are not viewed nutritionally. They are more known as adjuncts used in small quantities for flavouring numerous foods, as anti-oxidants, as preservatives, or in medicines or for the manufacture of essential oils for perfumery, essences and medicines etc.

Table 179. Chemical composition of Ajowan ground spice per 100 gm

Constituents	Amount
Moisture (g)	7.4
Carbohydrate (g)	24.1
Protein (g)	17.1
Crude fibre (g)	21.2
Fat (Ether extract) (g)	18.1
Mineral matter (Total ash) (g)	7.9
Calcium (g)	1.5
Phosphorus (g)	0.30
Iron (mg)	14.6
Sodium (mg)	56.0
Thiamine (mg)	0.21
Ribiflavin (mg)	0.18
Nicotinic acid (mg)	2.1
Carotene (mg)	71.0
Caloric value	363

32.1.ORIGIN AND DISTRIBUTION

Ajowan is native to Egypt and is cultivated in Mediterranean region and South Asian countries such as Iran, Iraq, Afghanistan, Pakistan, India, and even in Egypt and in Africa. In India, it is grown in the States of Andhra Pradesh, Bihar, Gujarat, Jharkhand, Madhya Pradesh, Maharashtra, Punjab, Rajasthan, Tamil Nadu, Uttar Pradesh, Uttaranchal and West Bengal.

32.2. CLIMATE

Ajowan is cultivated extensively as a cold season crop in the plains and as a summer crop in the hills. It requires a warm and long frost–free growing season and warm weather during seed development. The crop has moderate tolerance to drought and possesses wider climatic adaptability. In India, it is grown under both rainfed and irrigated cultivation system.

32.3. SOIL

Ajowan can be grown in any type of soil ranging from loamy to sandy loam and even in black soils. But it grows well in humus–rich loam and clayey loam soils.

32.4. LAND PREPARATION

Land should be brought to fine tilth by shallow ploughing 4–5 times with tractor or power tiller giving sufficient intervals between two ploughings. Land should be levelled and divided into small plots and channels to facilitate proper irrigation and intercultural operations. The manures should be applied at the time of land preparation and mixed well with soil by ploughing followed by planking.

32.5. SOWING OF SEEDS

1. Time of sowing: September–October

2. Method of sowing

The seeds should be sown, broadcasted or drilled in rows 45 cm apart at a depth of 1 cm. The plant to plant spacing should be maintained at 20–30 cm. The seeds should be sown in well–prepared seed beds because seeds are small and have a low germination rate of 60–70%.

3. Seed treatment

Seeds should be healthy and free from diseases. The seeds should be treated with any one of the following fungicides to protect the crop from seed–borne or to some extent soil–borne diseases also:

Table 180. Fungicides used for Seed treatment of Ajowan

Sr. No.	Name of fungicide/bio-fungicide	Amount (per litre of water)
1.	Carbendazim 50 WP	2.0–3.0 g
2.	M.E.M.C 6 WP	1.0 g
3.	Captan 50 WP	2.0–5.0 g
4.	Mancozeb 75 WP	2.5 g
5.	Trichoderma viridi	2.1 g

Seed rate

- Broadcasting - 4.0 kg/ha
- Drilling (i.e., sowing in lines) - 1.5 kg/ha

32.6. MANURING

Manures: Compost or FYM: 10.0–15.0 t /ha

Fertilizers

Nitrogen (N)	: 80.0–100.0 kg/ha
Phosphorus (P_2O_5)	: 30.0–50.0 kg/ha
Potassium (K_2O)	: 30.0–50.0 kg/ha
Sulphur (S)	: 50.0 kg/ha

The half dose of nitrogen, full quantity of phosphorus, potassium and sulphur should be added as a basal dose at the time of land preparation, and the remaining dose of nitrogen should be applied in two equal parts at an interval of 30 and 60 days after sowing as top dressing. Krishnamoorthy et al (2000) have reported a significantly higher seed yield of 2,050 kg/ha, when phosphorus 50 kg/ha and nitrogen 100 kg/ha were applied.

32.7. INTERCULTURAL OPERATION

1. Weeding

Weeds pose problem in the growth and development of the crop, so the crop should be kept free from weeds by weeding at regular intervals.

2. Thinning

The crop should be thinned at a spacing of 35 cm within 30 DAS. Thinning should be carried out when the plants attain a height of 20–25 cm.

32.8. IRRIGATION

The crop should be provided irrigation as per the need of the crop. Five to six irrigations are required depending on the climate and soil type.

32.9. HARVESTING

The small white flowers bloom in November and December in the plains and in mid–summer in the hills. The harvesting is usually done from February to May. Flower production ceases when the seeds start maturing and becomes greyish–brown in colour. The crop should be harvested at 120–140 DAS, when the flower heads turn brown. The mature umbels can be harvested with sickle manually. The harvested crop should be tied in small bundles and staked for drying, keeping upside down. The dried crop should then be threshed on a clean floor and winnowed to separate the clean seeds.

32.10. YIELD

400–600 kg/ha under a rainfed farming system and 1200–2000 kg/ha under irrigated conditions.

32.11. VARIETIES

Gujarat ajowan-1, Lam Sel-1, Lam Sel-2, S-25, S-47, S-84, S-95, IC-3743 etc.

32.12. PLANT PROTECTION

Insect pests

Insects do not cause much damage to the crop.

Diseases

1. Collar root rot

C.O.: *Sclerotium rolfsii*

Collar rot is a most serious disease of this crop.

Symptoms

The most obvious symptom found on an infected tree is a partial or complete girdling of the trunk. The diseased tissue at first appears orange and reddish-brown, then becomes browner as the rot and decay advances. The disease spread is most prevalent in areas that have poor drainage or standing water.

Control measures

- Proper drainage helps in reducing the incidence of collar rot.
- The disease can be controlled by spraying the crop with any one of the fungicides given in Table 181.

Table 181. Fungicides used for controlling the fungal diseases of mint

Sr. No.	Name of fungicide	Amount (per litre of water)
1.	Carbendazim 50 WP	1.0 g
2.	Chlorothalonil 75 WP	2.0 g
3.	Mancozeb 75 WP	2.5 g
4.	Captan 50 WP	2.5 g
5.	Copper oxychloride 50 WP	5.0 g

2. Powdery mildew

C.O.: *Erysiphe polygoni*

Powdery mildew appears at flowering stage in cloudy weather during the month of February-March

Symptoms

The symptoms of powdery mildew disease first appear on leaves and then other green parts of the plant. In initial stages, small, white circular powdery spots appear on the upper surface of leaves, which gradually cover the entire leaf, stem, petiole and fruits. In advanced stages of the disease, the leaf appears as if covered by talcum like powder. The plant assumes a greyish appearance and then turns yellow and fruits are shed profusely. The disease–affected plants lose their vigour and remain barren, and the yield is very much reduced.

Control measures

- The crop residues should be destroyed by burning as the fungus perpetuates in the debris of the plant.
- Late sowing should be avoided.
- The disease can be controlled by dusting the crop with sulphur 80 WP @ 3.0 kg per hectare or spraying the crop with any one of the fungicides given in Table 182, starting just on the appearance of the disease, at fortnight interval.

Table 182. Fungicides -Powdery mildew of Chilli

Sr. No.	Name of fungicide	Amount (per litre of water)
1.	Tridemorph 80 EC	0.5 ml
2.	Sulphur 80 WP	2.0–3.0 g
3.	Hexaconazol 5 EC	1.5 ml
4.	Myclobutanil 10 WP	0.5 g

33 TAMARIND

Scientific name - *Tamarindus indica*

Family - Fabaceae

Commercial part – Fruit

Indian names - *Bengali* : Tetul; *Hindi* : Imli; *Oriya* : Tentuli; *Karnataka* : Hunasae hannu; *Assamese* : Teteli; *Gujarati* : Amli; *Rajasthani* : Aamli; *Marathi* : Hunase; *Telugu* : Chintachettu (tree); *Malayalam* : Vaalanpuli; *Sanskrit* : Thinthrinee.

Tamarind (from Arabic: romanized Tamar hind) or commonly known as 'Indian date', is a tree in the family Fabaceae indigenous to tropical Africa. The genus Tamarindus is a monotypic taxon having only a single species. Tamarind (*Tamarindus indica*) (from Arabic *tamar hind*, "Indian date") is a leguminous tree in the family Fabaceae. The genus *Tamarindus* is a monotypic taxon, having only a single species. The Tamarind is a long–lived, medium-growing, bushy tree, which attains a maximum crown height of 12–18 m.

Table 183. Nutritive value of Tamarind per 100 g of edible portion

	Pulp (ripe) *	Leaves (young)	Flowers
Calories	115	–	–
Moisture	28.2-52 g	70.5 g	80 g
Protein	3.10 g	5.8 g	0.45 g
Fat	0.1 g	2.1 g	1.54 g
Fiber	5.6 g	1.9 g	1.5 g
Carbohydrates	67.4 g	18.2 g	–
Invert Sugars	30-41 g	–	–
(70% glucose; 30% fructose)	–	–	–

Ash	2.9 g	1.5 g	0.72 g
Calcium	35-170 mg	101 mg	35.5 mg
Magnesium	–	71 mg	–
Phosphorus	54-110 mg	140 mg	45.6 mg
Iron	1.3-10.9 mg	5.2 mg	1.5 mg
Copper	–	2.09 mg	–
Chlorine	–	94 mg	–
Sulfur	–	63 mg	–
Sodium	24 mg	–	–
Potassium	375 mg	–	–
Vitamin A	15 I.U.	250 mcg	0.31 mg
Thiamine	0.16 mg	0.24 mg	0.072 mg
Riboflavin	0.07 mg	0.17 mg	0.148 mg
Niacin	0.6-0.7 mg	4.1 mg	1.14 mg
Ascorbic Acid	0.7-3.0 mg	3.0 mg	13.8 mg
Oxalic Acid	–	196 mg	–
Tartaric Acid	8-23.8 mg	–	–
Oxalic Acid	trace only	–	–

Fig 61. (1) Tamarind twig with fruit (2) Tamarind fruit

India is the only producer of Tamarind on a commercial scale. A large part of India's production of Tamarind is exported to West Asia, Europe and America, where it is used for food specialties like Worcestershire sauce. It is extensively used in Asian cooking and is often used to make juices, soups, chutneys and other savoury dishes. Tamarind, an evergreen tree and native of East Africa, is now grown extensively in India, South East Asia and the West Indies. The Tamarind tree produces edible, pod–like fruit which are used extensively in cuisines around the world. The fruit pulp is edible. The ripened fruit is considered more palatable, as it becomes sweeter and less sour (acidic) as it matures.

33.1. CLIMATE

The Tamarind is well–adapted to humid tropical area with seasonally high rainfall. Dry weather is important during the period of fruit development. It prefers warm climate and grows well in semi–arid tropical and sub–tropical regions. It is very sensitive to frost. Tamarind trees are found growing naturally up to an elevation of 500 m from the MSL.

33.2. SOIL

Tamarind can grow well in any kind of soil provided there is sufficient moisture in the soil. Tamarind tree grows well in deep, well–drained soils, which are slightly acidic. Trees cannot tolerate heavy cold and wet soils, but are tolerant of salt spray and can be planted fairly near the seashore.

33.3. PROPAGATION

Tamarind is propagated through seeds and vegetative propagation. Seed propagation is common in Tamarind. Seeds, collected from fruits of healthy and vigorous trees, should be sown in well–prepared nursery beds at 15 x 10 cm distance and at 9 cm depth. Seedlings can also be raised in polythene bags. The seeds, soaked in water for 24 hours before sowing, germinate easily. Generally, seeds germinate 15–20 DAS in bed. The seedlings should be carefully removed from the nursery so that their roots are not damaged and then should be planted in the field.

Among vegetative propagations, budding, cutting and grafting are quite successful. Cuttings root easily in rainy season or in glass house.

The seedlings should be planted at a distance of 12 x 12 m in a field or 10 m apart as avenue tree on farm boundary in pits of 1 x 1 x 1 m size that are prepared in summer, refilled with soil along with 50 kg compost. Planting should be done in the beginning of monsoon. Trees grow easily from seeds and usually take 4–5 yr before they begin flowering and fruit production.

33.4. MANURING

Tamarind may not require nitrogenous fertilizers as it is a leguminous plant.

33.5. IRRIGATION

Young trees require adequate soil moisture until they become established, but mature trees do quite well without supplemental irrigation. Over–watering should be avoided as it results in soggy soils.

33.6. HARVESTING

Pods are harvested at different stages of ripeness according to how they are going to be used:

- **Immature green fruits** - Immature green fruits are usually harvested earlier for flavouring and making chutney.
- **Mature fruits** - Mature fruits are harvested later. Tamarind fruits mature in early summer. They are often gathered by shaking the tree and collecting the fruits that have fallen. The pods should be allowed to ripen on the tree until the outer shell is dry.

The tree begins to bear fruit at the age of 13–14 yr and continues to yield abundant crops for more than 60 yr. The flowers appear from June and July, and the pods ripen in the cold season. Mostly the harvesting of Tamarind takes place between 1st March to 15th April. Fruits harvested by the pulling pod away from the stalk. The fruits harvested are stored in neat and clean place. The harvested pod should be sun dried for getting best pulp.

33.7. PROCESSING

Tamarind should be deseeded. For the deseeding purpose, Tamarind should be sun–dried for two days so that the pulp remains intact.

33.8. YIELD

A fully–developed tree can give production of 200–250 kg fruits per annum.

33.9. VARIETIES

PKM-1, Urigam, Tumkur etc.

33.10. USES

1. **Wood** - Purplish brown hardwood is used for furniture, well construction, tent pegs, canoes, side planks, boats, cart shafts and axles, naves for wheels, toys, hubs, oil, sugar presses and tool handles. Wood is used as fuel as it has high calorific value (4850 k cal/kg) and also used for brick making; excellent for charcoal.

2. **Fruit and pulp** - Fruit and pulp product having economic value is sold in local markets in fresh or processed form with high export value. Fruits of sweet varieties are eaten fresh and sour fruits are used to prepare juice, jams, jellies and candy. The juice is sold commercially in polypacks and can also be used as vinegar to preserve food. Acidic pulp is used in curries, chutneys, sauces, ice cream, sherbets and toffees. The pulp can also be mixed with common salt to clean brass, copper and silver. Fruit pulp improves appetite and is a gentle laxative (medicinal).
3. **Seed** - Seeds are eaten roasted or boiled and are used in jams, jellies, confectionery and condiments. In some countries seeds are eaten alone or mixed with cereal powders, made into flour for bread and cake making. Seeds are a good substitute for fruit proteins. Seeds are used to manufacture Tamarind Kernel Powder (TKP), an adhesive used in textile and paper sizing, and in the food industry to thicken and stabilize. TKP is also used as filler for adhesives in the plywood industry, where seeds, when ground, boiled and mixed with gum, produce strong wood cement. Oil extracted from the seed can be used for making varnishes, paints and lighting oil. Seed is also a source of pectin which can be used in the food industry. Seed meal is a good livestock feed. Powdered seed is used to treat chronic diarrhea, jaundice and dysentery (medicinal).
4. **Leaf** - Leaves are edible, used to make curries, soups, salads and stews. Young leaves are used as a seasoning vegetable. Fodder relished by cattle and goats, also preferred by wild animals. Leaves contain tannin and can be used as an ink and in fixing dyes. Leaves are used in cardiac and blood sugar reducing medicines (medicinal). Leaves ground with salt are used to prevent throat infections, coughs, fever, intestinal worms and liver ailments (medicinal). Leaves may be applied to boils/abscesses to reduce inflammation (medicinal).
5. **Flowers and immature pods** - Flowers and immature pods are edible and can be used in salads and stews. Leaves may be used internally as a remedy to cure jaundice and externally to cure eye diseases and skin ulcers (medicinal).
6. **Twigs and bark** - Twigs are used in the tanning industry for tanning hides. The bark can be burnt to make ink. Twigs and bark can be used to make dyes. Twigs are used as chewsticks. Bark is used as an astringent for the treatment of diarrhoea, as a cure for asthma and acts as a digestive aid (medicinal).
7. **Root** - Roots are used as component in medicines for leprosy and dysentery (medicinal).

34 HORSERADISH

Scientific name - *Armoracia rusticana*

Syn. - *Cochlearia armoracia*

Family - Brassicaceae

Commercial part - Root

Horseradish, a perennial plant, is probably native to south–eastern Europe and western Asia. It is now popular around the world. It grows up to 1.5 m (5 ft) tall, and is mainly cultivated for its large, white, tapered root. Horseradish is planted from tuberous roots rather than seeds. Horseradish is a flavourful herb that has been used for centuries to enhance the flavour of food, to aid in digestion and to keep people healthy.

Fig 62. Horsegram

Fig 62 (a) Horsegram seed

35.1. CLIMATE

Horseradish is grown in temperate climates around the world, primarily as an annual, and to a limited extent, as a perennial crop.

35.2. SOIL

Horseradish grows best on rich, moist, deep friable loam or sandy loam soils rich in organic matter. It tolerates a wide soil pH range of 5.5–6.8.

35.3. PLANTING

1. Time of planting

Horseradish is propagated from root cuttings or crowns and planted in the spring time.

2. Seed treatment

Seeds should be healthy and free from diseases. The seeds should be treated with any one of the fungicides given in Table 184 to protect the crop from seed–borne or to some extent soil–borne diseases also.

Table 184. Fungicides used for seed treatment of Horseradish

Sr. No.	Name of fungicide/bio-fungicide	Amount (per litre of water)
1.	Carbendazim 50 WP	2.0–3.0 g
2.	M.E.M.C. 6 WP	1.0 g
3.	Captan 50 WP	2.0–5.0 g
4.	Mancozeb 75 WP	2.5 g
5.	Trichoderma viridi	4.0 g

3. Seed rate

A range of 15,000–22,000 sets per hectare is needed, depending on the row spacing. Rows can be spaced 3–4 ft (90–120 cm) with 20–24 inches (50–60 cm) between plants in the row.

4. Method of planting

Sets are manually planted by laying them on the soil surface horizontally at a slight angle, with the head–end placed slightly higher.

35.4. MANURING

Manures

Compost or FYM: 12–20 t/ha

Fertilizers

Nitrogen (N)	: 112.50–225.0 kg/ha
Phosphorus (P_2O_5)	: 112.50–168.75 kg/ha
Potassium (K_2O)	: 112.50–168.75 kg/ha
Boron (B)	: 2.25–3.5 kg/ha
Sulphur (S)	: 33.75–56.25 kg/ha

The total amount of manures and fertilizers should be applied at the time of land preparation and mixed with the soil by cross ploughing followed by planking.

35.5. HARVESTING

Horseradish should be harvested by digging a trench 12–24 inches (30-60 cm) deep along one side of the row.

Harvest should begin once the tops are frozen back. The harvested root mass should be brought into a sorting area. The roots can be shaken clean by hand or washed with a hand water spray.

35.6. YIELD

75.0–90.0 q/ha

35.7. PLANT PROTECTION

Diseases

Cercospora leaf spot (C.O.: *Cercospora armoraciae*)

Ramularia leaf spot (C.O.: *Ramularia armoraciae*)

Bacterial leaf spot (C.O.: *Phytomonas campestre armoraciae*)

Insect pests

Brittle Root (*Spiroplasma citri*)

Flea beetles (*Horseradish flea beetle*).

35 ROSEMARY

Scientific name - *Rosmarinus officinalis*

Family - Lamiaceae

Indian names - *Hindi* : Rusmari; *Bengali* : Rosemary; *Sanskrit* : Rusmari

Rosemary is a woody, perennial herb with fragrant, evergreen, needle–like leaves and white, pink, purple, or blue–coloured flowers. The name 'Rosemary' derives from the Latin word 'dew' (*ros*) and 'sea' (*marinus*), or 'dew of the sea'. Rosemary is the native of Mediterranean region and is cultivated in Europe and California in US. It is also grown in Algeria, China, Middle East, Morocco, Russia, Romania, Serbia, Tunisia, Turkey, and to a limited extent in other parts of the world. It is a perennial crop and is commercially viable up to 12 years.

35.1. CLIMATE

Rosemary requires cool winter and mild summer below 30°C. The temperate climatic zones ranging from 900– 2,500 m above MSL are suitable for Rosemary cultivation.

35.2. SOIL

Rosemary requires a well-drained loamy soil with a pH of 5.5–7.0. When the pH is below 5.0, dolomite @ 2.5 t/ha should be applied and mixed well in the soil.

35.3. LAND PREPARATION

The land is brought to fine tilth by cross ploughing with tractor or power tiller followed by planking. At the time of last ploughing, 10.0 t of well decomposed FYM and1.25 t of neem cakes per hectare should be applied and mixed well with the soil. A bed of 30 cm height, 1.5 m width and a length of convenient size is prepared. At the time of planting, 5.0 kg of *Azospirillum* and 5.0 kg *Phosphobacterium* should be applied to the soil and mixed well.

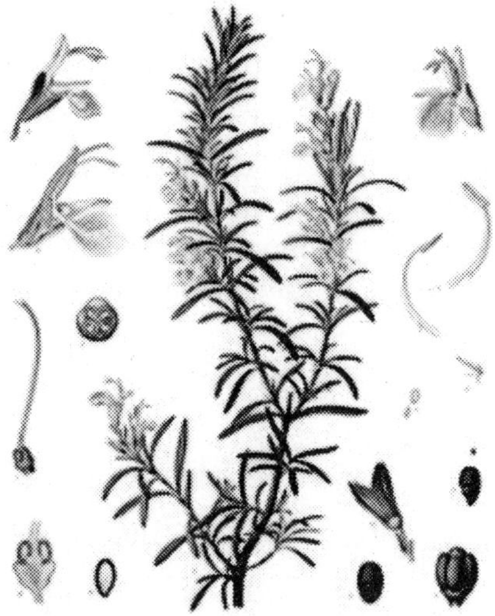

Fig 63. Rosemary

35.4. PROPAGATION

Rosemary is propagated by seeds, cuttings and layers, and by division of roots. Rosemary rooted cuttings can be planted during the months of June–July and September–October under rainfed conditions.

The rooted cuttings of Rosemary should be planted at a spacing of 45 x 45 cm. The central shoot should be removed six months after planting to promote lateral shoots.

35.5. MANURING

Manures

Compost: 10 t/ha

Or

Vermicompost: 5.0 t/ha.

Neem cake: 1.25 t/ha

35.6. AFTER CULTIVATION

The land should be kept clean by regular weeding. First hoeing and weeding should be done one month after planting. 4–5 weeding should be done in a year.

The crop can be grown under rainfed conditions as a dry farming crop.

35.7. HARVESTING

Rosemary leaves are harvested with the onset of flowering. The flowering tops measuring 30–35 cm long with leaves are harvested with a sickle.

The leaves should be washed thrice for removing various dirt and soil particles. The leaves should be dried under shade. Rosemary has many uses besides culinary. It is used in medicines or as a dye in cosmetics. It is an aromatic plant, ornamental plant in landscape and a houseplant.

35.8. YIELD

Dried leaves: 2.5 t/ha per year

35.9. VARIETIES

Albus, Arp, Aureus, Benenden Blue, Blue Boy, Golden Rain, Gold Dust, Irene, Lockwood de Forest, Ken Taylor, Majorica Pink, Miss Jessop's Upright, Pinkie, Prostratus, Pyramidalis (or Erectus), Remembrance (or Gallipoli), Roseus, Salem, Severn Sea, Tuscan Blue and Wilma's Gold.

36

THYME

Scientific name - *Thymus vulgaris*

Family - Labiatae

Indian names - *Hindi* : Banajwain; *Malayalam* : Thottathulasi; *Punjabi* : Marizha, Masho, Rangsbur; *Urdu* : Hasha

Thyme is one of the best–known and most widely–used culinary herbs. It is quite easy to grow and is commonly found as a decorative as well as a functional plant in many home gardens.

Thyme is a low–growing, woody, perennial plant. Thyme plants should be spaced 18–24 inches (45–60cm) apart. Thyme grows well in a light, well–drained, poor to fertile soil. Thyme grows in a pH range between 6.5 (neutral) and 8.5 (alkaline) with a preferred pH range between 6.5 and 7.0. Thyme can be propagated in a variety of ways such as seed, root division and from cuttings.

Thyme is harvested in mid–summer, just prior to flowering. Thyme is used both fresh and dried.

Thyme has been used since ancient times for its culinary, aromatic and medicinal properties.

37 SAGE

Scientific name - *Salvia officinalis*

Family - Lamiaceae

Commercial part - Leaf

Indian names - *Hindi* : Salvia, Sefakus; *Malayalam* : Salvi tulasi; *Bengali* : Bui tulasi; *Panjabi* : Sathi

Sage is native to the Mediterranean and cultivated all over the world. It grows wild in the Dalmatian region of Yugoslavia. It is cultivated in Yugoslavia, Italy, Albania, Turkey, Portugal, Spain, Cyprus, England, Canada and USA. In India, it is sparingly cultivated in Jammu. Common Sage, or *Salvia officinalis*, is the basic Sage that has been cultivated throughout Europe for culinary and medical uses since ancient times. The name 'Sage' means 'safe' or 'cure' referring to the plant's healing powers. For thousands of years, Sage is prized for its soft, textured leaves both in the kitchen and the garden.

Sage is grown for its aromatic leaves. Sage thrives well in rich clayey and loamy soil. Sage prefers a warm, sunny location although it does not enjoy extreme heat. It is not particular about soil, except that it is well–drained. Sage grows well in fertile garden soil. Sage plants can be propagated from seed, root cuttings or transplants. Sage seeds need to be sown while fresh.

Pruning after flowering keeps plants attractive and prevents them from getting too woody.

The individual leaves should be harvested as needed. Leaves can also be dried and stored for future use. Sage is best used fresh, although leaves frozen in a plastic bag are an excellent alternative. Sage is used for flavouring meat and fish dishes and for poultry stuffing. Fresh Sage leaves are used in salads and sandwiches.

Fig 64. Sage

Sage is used as a mild tonic, astringent and carminative. It is diaphoretic and anti-pyretic. Sage oil is used in perfumes as a deodorant. Sage and Sage oil exhibit anti-oxidant properties.

38 TARRAGON

Scientific name - *Artemisia dracunculus*

Family - Asteraceae

Parts Used - Leaves

Tarragon, (*Artemisia dracunculus*) is a species of perennial herb in the family Asteraceae. Tarragon, believed to be a native of southern Russia, Siberia and Tartary has been cultivated for around 600 yr. French tarragon is grown for its distinctively flavoured leaves. Tarragon is a perennial herb cultivated for the use of its aromatic leaves in seasoning, salads, etc., and in the preparation of Tarragon vinegar.

Table 185. Composition dried leaves of Tarragon

Moisture	4.7%
Protein	24.2%
Fat	7.3%
Fiber	6.8%
Carbohydrates	44.7%
Total ash	12.3%
Calcium	1.3%
Phosphorus	0.31%
Iron	0.034%
Sodium	0.07%
Potassium	3.2%
Vitamin A (carotene)	175 I.U./100 g
VitaminB1	0.2mg/100 g
VitaminB2	1.34 g/100 g

Niacin	8.9 mg/100 g
Vitamin C (ascorbic acid)	12 mg/100 g
Calorific value	365 calories

Fig 65. Russian Tarragon

It grows well in a rich loamy soil that holds moisture, but drains well. Mulching is beneficial to this end. Tarragon is only propagated by cuttings, layers and divisions, and planted at least 45 cm apart. Tarragon is one of the four fines herbs of French cooking, and is particularly suitable for chicken, fish and egg dishes. Tarragon enhances the flavour of many foods. The leaves are picked anytime. A main crop occurs in late summer (June–October). It grows well in full Sun or partial shade, and needs well–drained soil. It is propagated from stem tip or root cuttings, or from crown divisions. The plants should be planted at about 60-90 cm ft apart to give them room to spread.

Harvesting should be done regularly or plants should be pruned to a height of 60 cm to prevent flowering and to keep plants neat and clean.

REFERENCES

Atal EK and Kapur BM 1982, Cultivation and Utilization of Medicinal and Aromatic plants, New Delhi: CSIR

Chadha KL and Rethinam P 1994, Advances in Horticulture (Volume 9), Plantation and Spice Crops (Part 1 and 2), New Delhi: Malhotra Publishing House

Das PC 2006, *Vegetable Crops of India (Second Ed.),* Ludhiana: Kalyani Publishers

Weiss EA 2002, *Spice crops*, Wallingford: CABI Publishing

Farooqi AA, Sreeramu BS and Srinivasappa KK 2005, *Cultivation of Spices Crops*, Hyderabad: Universities Press (India) Pvt. Ltd.

Godara SL, Kapoor BBS and Rathore BS 1976, Cultivation of Spice Crops, Bikaner: Madhu Publications

George CK, Anandan A and Chapman K 2005, *Pepper Production Guide for Asia and the Pacific,* Indonesia: International Pepper Community

Kumar N, JBM Md. Abdul Khander, Rangaswami P and Iruläppan I 2001, *Introduction to Spices, Plantation Crops, Medicinal and Aromatic Plants*, New Delhi: Oxford & IBH Publishing Co. Pvt. Ltd.

Leela NK 2008, Cinnamon and Cassia, in *Chemistry of Spices* (eds) A Villupanoor, BC Parthasarathy, and JT Zachariah, Wallingford: CABI Publishing

Peter KV 2007, *Spices, Horticulture Science Series, Vol. 5*, New Delhi: New India Publishing Agency

Ravindran PN 2006, Advances In Spices Research: History And Achievements Of Spices Research In India Since Independence, Jodhpur: Agrobios (India)

Pradeep kumar T, Jyothibhaskar BS and Satheesan KN, 2008, *Management of Horticultural Crops*, New Delhi: New India Publishing Agency

Purseglove JWEG, Brown Green CL and Robbins SRJ 1980, *Spices (Vol. I & II),* New Delhi: Academic press

Pruthi TS 1976, Minor Spices and Condiment Crop Management and Postharvest Technology, New Delhi: I.C.A.R.

Ramaswamy N 1994, *Fruits, Spices and Plantation Crops,* Madras: Camino Crafts

Reddy KB and Reddy IVS 2008, Production Technology of Spices, Aromatic, Medicinal and Plantation Crops (Hort-282), Hyderabad: Acharya N.G. Agricultural University

Shanmugavelu KG and Madhava Rao VN 1977, Spices and plantation crops, Chennai: Popular Book Depot

Thankamani CK, Srinivasan V, Dinesh R, Santhosh JE and Rajeep P 2009, *Black Pepper (Extension Pamplet)*, Calicut: Indian Institute of Spice Research.

APPENDICES

APPENDIX 1

1. Land and its measurement

1 bigha	20 Katta
1 bigha	33 cent (40 cent in some area)
1 acre	3 bigha or 2.5 bigha
1 acre	100 cent or 43, 560 sq ft or 0.40 hectare
1 hectare	2.471 acre or 10, 000 sq m

2. Weight: Its units and conversion

1 ounce	28.3495 gm
1 pound	0.454 kg
1 tola	11.6638 gm
1 chhatak	58.318 gm
1 maund (40 sers)	37.3242 kg or 0.37 quintal
1 metric ton	1,000 kg
1 quintal	100 kg or 2.679 mound

3. Length: Its unit and conversion

1 foot	12 inches
1 yard	3 feet or 0.9 m
1 inch	2.54 cm or 25.40 mm
1 feet	30.48 cm
1 metre (m)	39.3701 inches
1 kilometre (km)	1093.61 yard or 0.62137 mile

1 sq cm	0.155 sq inch
1 sqft	92.903 sq cm
1 sqyard	0.48 sq m
1 sqm	10, 000 sq cm

4. Liquid: Its units and conversion

1 gallon	4.543 litre
1 litre	0.22 gallon

5. Weight per acre into kilogram per hectare

1maund per acre	0.9223 kg per hectare
1 pound per acre	1.1209 kg per hectare
1 ounce per acre	70.05 kg per hectare
1 cwt per acre	125.6 kg per hectare

Appendix 2. Nutrient content in manures

Manures	Nutrient percentage		
	N	P_2O_5	K_2O
A. Bulky organic manures			
Cow dung (Fresh)	0.3–0.4	0.1–0.2	0.1–0.3
Horse dung (Fresh)	0.4–0.5	0.3–0.4	0.3–0.4
Sheep dung (Fresh)	0.5–0.7	0.4–0.6	0.3–1.0
Night soil (Fresh)	1.0–1.6	0.8–1.2	0.2–0.6
Poultry manure (Fresh)	1.0–1.8	1.4–1.8	0.8–0.8
Sewage and sludge (Dry)	2.0–5.0	1.0–5.0	0.2–0.5
Sewage and sludge, activated dry	4.0–7.0	2.1–4.2	0.5–0.7
Urine, Cattle	0.9–1.2	Trace	0.5–1.0
Urine, Horse	1.2–1.5	Trace	1.3–1.5
Urine, Human	0.5–1.0	0.1–0.2	0.2–0.3
Urine, Sheep	1.5–1.7	Trace	1.8–2.0
Farm Yard Manure	0.5–1.5	0.4–0.8	0.5–1.9
Compost (Rural), dry	0.4–0.8	0.3–0.6	0.7–1.0
Compost (Urban), dry	1.0–2.0	1.0	1.5
Water hyacinth compost	2.0–3.0	1.0–2.0	3.0–4.0
Green manures (Various, average)	0.5–0.7	0.1–0.2	0.6–0.8
(i) Dhaincha (Sesbaniaaculeata)	0.62	0.15	0.58
(ii) Cowpea (Vignacatjang)	0.75	0.150	0.58
(iii) Sannhemp (Crotolariajuncea)	0.75	0.15	0.58
B. Manures of animal origin			
Dried blood	10.0–12.0	1.0–1.5	1.0
Fish manure	4.0–10.0	3.0–9.00	0.3–1.5
Bone meal (Raw)	2.0–4.0	20.0–25.0	–
Bone meal (Steamed)	1.0–2.0	25.0–30.0	–
C. Wood ash			
Ash, household	05–1.9	1.6–4.2	2.3–12.0
Ash, Coal	0.73	0.45	0.53
Ash, Babul	0.1	0.8–1.3	1.5–3.1

D. Crop residue			
Paddy husk	0.3–0.5	0.2–0.3	0.3–0.5
Straw	0.36	0.08	0.71
Groundnut shell and stem	1.6–1.8	0.3–0.5	1.1–1.7
E.Oil cakes			
1. Edible oil cakes			
Mustard cake	5.1–5.2	1.8–1.9	1.1–1.2
Linseed cake	5.5–5.6	1.4–1.5	1.2–1.3
Til or Sesame cake	6.2–6.3	2.0–2.1.0	1.2–1.3
Coconut cake	3.0–3.2	1.9–2.0	1.7–1.8
Safflower cake (Decorticated)	7.9	2.2	1.9
Cotton seed cake (Decorticated)	6.4	2.9	2.2
Groundnut cake	7.0–7.3	1.5–1.6	1.3–1.4
2. Non–edible cake			
Castor cake	4.3	1.8	1.3
Cotton seed cake (Undecorticated)	3.9–4.0	1.8–1.9	1.6–1.7
Karanja or honge cake	3.9.5	0.9	1.2
Mahua cake	2.5–2.9	0.8–0.9	1.8
Neem cake	5.2–5.3	1.0–1.1	1.4–1.5
Safflower cake (Undecorticated)	4.9	1.4	1.2

Appendix 3. Nutrients content in fertilizers

Name of fertilizer	Nutrient percentage		
	N	P_2O_5	K_2O
1.Nitrogenous fertilizers			
(i) Ammonium sulphate [$(NH_4)_2$ SO_4]	20.6–21.0	–	–
(ii) Urea [$CO(NH_2)]_2$	44.0–46.0	–	–
(iii) Ammonium chloride (NH_4Cl)	25.0	–	–
(iv) Ammonium sulphate nitrate [$(NH4)_2SO_4$ + NH_4NO_3]	26.0	–	–
(v) Calcium ammonium nitrate [$CaNH_4(NO_3)_3$]	25.0	–	–
(vi)Calcium nitrate [$Ca(NO_3)2$]	15.6–25.0	–	–
(vii) Sodium nitrate ($NaNO_3$)	16.0	–	–
(viii) Calcium cyanamide ($CaCN_2$)	21.0	–	–
(ix) Anhydrous ammonia	80.0–82.0	–	–
(x) Ammonium nitrate (NH_4NO_3)	32.0–35.0	–	–
2.Phosphatic fertilizers			
(i) Super phosphate (Single) [$Ca(H_2PO_4)2.H_2O$]	–	16.0–20.0	–
(ii) Super phosphate (Double) [$2Ca(H_2PO_4)2.2H_2O$]	–	30.0–35.0	–
(iii) Super phosphate (Triple) [$3Ca(H_2PO_4)2.H_2O$]	–	45.0–50.0	–
(iv) Basic slag	–	3.0–8.0	–
(v) Rock phosphate			
(a) Mussori	–	23.0–24.0	–
(b) Purilia	–	23.0	–
3.Potassic fertilizers:			
(i) Muriate of potash or Potassium chloride (KCl)	–	–	50.0–60.0
(ii) Potassium sulphate (K_2SO_4)	–	–	48.0–52.0
(iii) Potassium phosphate (K_3PO_4)	–	–	30.0–50.0
4.Compound fertilizers:			
(i) Diammoniumphosphate	18.0	18.0	–
(ii)Urea ammonium phosphate (Gromor)	28.0	28.0	–
(iii)Monoammonium phosphate	11.0	11.0	–
(iv)Ammonium phosphate sulphate	20.0	18.0	–

(v) Nitro–phosphate potash (Suphala 15: 15: 15)	15.0	15.0	15.0
(vi) Gromor 14: 35: 14	14.0	35.0	14.0
(vii) IFFCO/Paras 10: 26: 26	10.0	26.0	26.0

Appendix 4. Universities and their addresses

S.N.	Name, email and website	Address	Telephone/ Fax No.
1.	Acharya NG Ranga Agricultural University Website: http: //www.angrau.net Email: angrau_vc@yahoo.com, raghuvardhanreddy_s@rediffmail.com	Administrative office, Rajendra Nagar, Hyderabad–500030, Andhra Pradesh	040–24015035, 24013095 Fax: 040–24015031
2.	Anand Agricultural University Website: http: //www.aau.in Email: vc@aau.in, vc_aau@yahoo.com	Anand 388110, Gujarat	02692–261273 Fax: 02692–261520
3.	Assam Agricultural University Website: http: //www.aau.ac.in Email: vc@aau.ac.in, kmbujarbaruah@rediffmail.com	Jorhat 785013, Assam	0376–2340013 Fax: 0376–2340001
4.	Bidhan Chandra Krishi Viswavidyalaya Website: http: //www.bckv.edu.in Email: bckvvc@gmail.com,	Mohanpur, Nadia–741252, West Bengal	033–25879772, 03473–222666 Fax: 03473–222275
5.	Bihar Agricultural University Website: http: //www.bausabour.ac.in Email: vcbausabour@gmail.com	Sabour, Bhagalpur 813210, Bihar	0641–2452606 Fax: 0641–2452604
6.	Birsa Agricultural University Website: http: //www.baujharkhand.org Email: vc_bau@rediffmail.com	Kanke, Ranchi–834006, Jharkhand	0651–2450500 Fax: 0651–2450850
7.	Central Agricultural University Website: http: //www.cau.org.in Email: snpuri04@yahoo.co.in, snpuri@rediffmail.com	P.O. Box 23, Imphal–795004, Manipur	0385–2415933 Fax: 0385–2410414
8.	Chandra Shekhar Azad University of Agriculture & Technology Website: http: //www.csauk.ac.in Email: vc@csauk.ac.in	Kanpur–208002, Uttar Pradesh	0512–2534155 Fax: 0512–2533808
9.	Chaudhary Charan Singh Haryana Agricultural University Website: http: //www.hau.ernet.in Email: vc@hau.ernet.in	Hisar–125004, Haryana	01662–231640, 284301 Fax: 01662–234952

10.	CSK Himachal Pradesh Krishi Vishvavidyalaya Website: http: //www.hillagric.ac.in Email: vc@hillagric.ac.in	Palampur–176062, Himachal Pradesh	01894–230521 Fax: 01894–230465
11.	Dr. Balasaheb Sawant Konkan Krishi Vidyapeeth Website: www.dbskkv.org Email: vcbskkv@yahoo.co.in	Dapoli, Dist, Ratnagiri – 415712, Maharashtra	02358–282064 Fax: 02358–282074
12.	Dr. Panjabrao Deshmukh Krishi Vidyapeeth Website: http: //www.pdkv.ac.in Email: vc@pdkv.ac.in	Krishinagar, Akola–444104, Maharashtra	0724–2258365 Fax: 0724–2258219
13.	Dr. Yashwant Singh Parmar University of Horticulture & Forestry Website: http: //www.yspuniversity.ac.in Email: vc@yspuniversity.ac.in, vcuhf@yahoo.com	Solan, Nauni–173230, Himachal Pradesh	01792–252363 Fax: 01792–252242
14.	Dr YSR Horticultural University Website: http: //www.drysrhu.edu.in Email: vc@drysrhu.edu.in	Administrative office, Venkataramanna-gudem, PB No. 7, West Godavari Dist, Tadepalligudem–534101 AP	08818–284311 Fax: 08818–284223
15.	Govind Ballabh Pant University of Agriculture & Technology Website: http: //www.gbpuat.ac.in Email: vcgbpuat@gmail.com	Pantnagar–263145, Dist. Udham Singh, Nagar, Uttaranchal	05944–233333, 233500 Fax: 05944–233500
16.	Guru Angad Dev Veterinary and Animal Science University Website: http: //www.gadvasu.in Email: vijay_taneja@hotmail.com, vcgadvasu@gmail.com	Ludhiana – 141004, Punjab	0161–2553320, 2553360 Fax: 0161–2553340
17.	Indira Gandhi Krishi Vishwavidyalaya Website: www.igau.edu.in Email: vcigkv@gmail.com	Krishak Nagar, Raipur–492006, Chhattisgarh	0771–2443419 Fax: 0771–2442302, 2443121
18.	Jawaharlal Nehru Krishi Viswavidyalaya Website: http: //www.jnkvv.nic.in Email: gkalloo_jnkvv@yahoo.co.in	Krishi Nagar, Adhartal Jabalpur–482004, Madhya Pradesh	0761–2681706 Fax: 0761–2681389
19.	Junagadh Agricultural University Website: http: //www.jau.in Email: vc@jau.in	Univ. Bhavan, Motibagh Junagadh–362001, Gujarat	0285–2671784 Fax: 0285–2672004

20.	Karnataka Veterinary, Animal and Fisheries Sciences University Website: http: //www.kvafsu.kar.nic.in Email: vckvafsu@yahoo.co.in dekvafsub@yahoo.com, sskvafsu@yahoo.co.in	Nandinagar, PB No 6, BIDAR 585401, Karnataka	08482–245264 Fax: 08482–245107
21.	Kerala Agricultural University Website: http: //www.kau.edu Email: vc@kau.in, vicechancellorkau@gmail.com	Vellanikara, Trichur 680656, Kerala	0487–2371928, 2370034, 2438001 Fax: 0487–2370019
22.	Kerala University of Fisheries & Ocean Studies Website: http: //www.kau.edu Email: kurup424@gmail.com	Papangad, Kochi–682506 Kerala	0487–2370117, 2703781, 2700964
23.	Kerala Veterinary and Animal Sciences University Website: http: //www.kcasu.ac.in Email: vc@kvasu.ac.in, vc.vetuny@gmail.com	Liason Office, Directorate of Dairy Development, Pattom, Thiruvananthapuram, 695004, Kerala	0471–2550058 Fax: 0471–2550480
24.	Lala Lajpat Rai University of Veterinary & Animal Sciences Website: http: //www.kcasu.ac.in Email: vc@kvasu.ac.in, vc.vetuny@gmail.com	Hisar, Haryana	0172– 2543002 01662– 289550
25.	Madhya Pradesh Pashu Chikitsa Vigyan Vishwavidyalaya Website: http: //www.mppcvv.org Email: vcmppcvv@yahoo.in	South Civil Lines, Jabalpur–482001, Madhya Pradesh	0761–2620783, 2678007 Fax: 0761–2620783
26.	Maharana Pratap University of Agriculture & Technology Website: http: //www.mpuat.ac.in Email: vc@mpuat.ac.in	Udaipur, Rajasthan 313001	0294–2471101 Fax: 0294–2470682
27.	Maharashtra Animal Science & Fishery University Website: http: //www.mafsu.in Email: vcmatsu@gmail.com, cadaba_prasad@yahoo.co.in	Seminary Hills, Nagpur–440006, Maharashtra	0712–2511088 Fax: 0712–2511282
28.	Mahatma Phule Krishi Vidyapeeth Website: http: //mpkv.mah.nic.in Email: vcmpkv@rediffmail.com	Rahuri–413722, Maharashtra	02426–243208 Fax: 02426–243302
29.	Manyavar Shri Kanshiramji University of Agriculture and Technology Website: http: //www.mskjuat.edu.in/ Email: vc.mskjuat@gmail.com	Banda – 210001, Uttar Pradesh	05192–221605 Fax: 02426–243302

30.	Marathwada Agricultural University Website: http: //www.mkv2.mah.nic.in Email: vcmau@rediffmail.com	Parbhani–431402, Maharashtra	02452–223002 Fax: 02452–223582
31.	Narendra Deva University of Agriculture & Technology Website: http: //www.nduat.ernet.in Email: vc_nduat2010@yaho.co.in	Kumarganj, Faizabad – 224229, Uttar Pradesh	05270–262097, 262161 Fax: 05270–262097
32.	Navsari Agricultural University Website: http: //www.nau.in Email: vc_2004@yahoo.co.in	Navsari–396450 Gujarat	02673–283869 Fax: 02673–284254
33.	Orissa Univ. of Agriculture & Technology Website: http: //www.ouat.ac.in Email: ouat_dproy@yahoo.co.in, bsenapati1942@yahoo.com	Bhubaneshwar–751003, Orissa	0674–2397700 Fax: 0674–2397780
34.	Punjab Agricultural University Website: http: //www.pau.edu Email: vcpau@pau.edu	Ludhiana–141004, Punjab	0161–2401794 Fax: 0161–2402483
35.	Rajasthan University of Veterinary and Animal Sciences Website: http: //rajuvas.org Email: vcrajuvas@gmail.com	BijeyBhavan Place Complex (Pt Dean Dayal circle), Bikaner – 334006, Rajasthan	0151–2543419 Fax: 0151–2549348
36.	Rajendra Agricultural University Website: http: //www.pusavarsity.org.in Email: vcrau@sify.com	Pusa, Samastipur – 848125, Bihar	06274–240226 Fax: 06274–240255
37.	Rajmata Vijayraje Scindia Krishi Vishwavidyalaya Website: http: //www.rvskvv.nic.in Email: vcrvskvv@gmail.com	Race Course Road, Gwalior – 474002, Madhya Pradesh	0751–2467673 Fax: 0751–2467673
38.	Sardar Vallabhbhai Patel University of Agriculture and Technology Website: http: //www.svbpmeerut.ac.in Email: vc_agunivmeerut@yahoo.com	Modipuram, Meerut – 250110, Uttar Pradesh	0121–2888522, Fax: 0121–2888505
39.	Sardar krushinagar–Dantiwada Agricultural University Website: http: //www.sdau.edu.in Email: vc@sdau.edu.in	SardarKrushinagar, Dist. Banaskantha, Gujarat–385506	02748–278222 Fax: 02748–278261
40.	Sher–E–Kashmir University of Agricultural Sciences & Technology Website: http: //www.skuast.org Email: vc@skuast.org	Railway Road, Jammu 18009, Jammu & Kashmir	0191–2263714 Fax: 0191–2262073

41.	Sher–E–Kashmir University of Agricultural Sciences & Technology of Kashmir Website: http://www.skuastkashmir.ac.in Email: vc@skuastkashmir.ac.in, skuastkvc@gmail.com	Shalimar Campus, Shrinagar–191121, Jammu & Kashmir	0194–2464028, 2462159 Fax: 0194–2462160, 2461543
42.	Sri Venkateswara Veterinary University Website: http://www.svvu.edu.in Email: prabhakarvrao@yahoo.com	Admin. office, Regional Library Building, Tirupati–517502	0877–2248986 Fax: 0877–2249222
43.	Swami Keshwanand Rajasthan Agricultural University Website: http://www.raubikaner.org Email: vcrau@raubikaner.org	Bikaner–334006, Rajasthan	0151–2250443, 2250488 Fax: 0151–2250336
44.	Tamil Nadu Agricultural University Website: http://www.tnau.ac.in Email: vc@tnau.ac.in	Coimbatore–641003, Tamil Nadu	0422–2431788, 2431672 Fax: 0422–2431672
45.	Tamil Nadu Veterinary & Animal Science University Website: http://www.tanuvas.ac.in Email: vc@tanuvas.org.in, karanmgk@gmail.com	Chennai–600051, Tamil Nadu	044–25551574, 25551575 Fax: 0444–225551576
46.	University of Agricultural Sciences, Bangalore Website: http: //www.uasbangalore.edu.in Email: vc@uasbangalore.edu.in, knarayanagowda@yahoo.co.in	GKVK, Bengaluru–560065, Karnataka	080–23332442 Fax: 080–23330277
47.	University of Agricultural Sciences, Dharwad Website: http://www.uasd.edu Email: vc_uasd@rediffmail.com	Dharwad–580005, Karnataka	0836–2447783, Fax: 0836–2448349
48.	University of Horticultural Sciences Website: http://uhsbagalkot.edu.in/ Email: dandinbnm@gmail.com	Sector No 60, Navanagar, Bagalkot – 587102 Karnataka	08354 201310 Fax: 08354 235152
49.	University of Agricultural Sciences Website: http://www.uasraichur.edu.in Email: vcuasraichur10@rediffmail.com	PB 329, Raichur–584101 Karnataka	08532–221444 Fax: 08532–220444
50.	UP Pandit DeenDayal Upadhaya Pashu Chikitsa Vigyan VishwaVidhyalaya evam Go Anusandhan Sansthan Website: http://www.upvetuniv.edu.in Email: singhambika1945@gmail.com, duvasuvc@gmail.com	Mathura–281001, Uttar Pradesh	0565–2470199 Fax: 0565–2404819

51.	Uttarakhand University of Horticulture and Forestry Website: http://www.uuhf.ac.in/ Email: deanhortbharsar@gmail.com	Pauri Garhwal, Uttarakhand	09412051976
52.	Uttar Banga Krishi Viswavidyalaya Website: http://www.ubkv.ac.in Email: vcubkv@gmail.com, vcubkv@rediffmail.com	P.O. Pundibari, Dist. Coach Bihar–736165, West Bengal	03582–270141, 270013 Fax: 03582–270249
53.	West Bengal University of Animal & Fishery Sciences Website: http://www.wbuafscl.ac.n Email: wbuafs@wb.nic.in	68 KB Sarani, Kolkata–700037, West Bengal	033–25563450 Fax: 033–25571986

Appendix 5. (1) List of some pesticides/ pesticides formulations banned in India

Sr. No.	Pesticides banned for manufacture, import and use	Pesticide / Pesticide formulations banned for use but their manufacture is allowed for export	Pesticide formulations banned for import, manufacture and use	Pesticide withdrawn
1.	Aldrin	Nicotin Sulfate	Methomyl 24% L	Dalapon
2.	Benzene hexachloride	Captafol 80% Powder	Methomyl 12.5% L	Ferbam
3.	Calcium cyanide		Phosphamidon 85% SL	Formothion
4.	Chlordane		Carbofuron 50% SP	Nickel Chloride
5.	Copper acetoarsenite			Paradichloro-benzene (PDCB)
6	Bromochloropropane			Simazin
7.	Endrin			Warfarin
8.	Ethyl mercury chloride			
9.	Ethyl parathion			
10.	Heptachlor			
11.	Menazone			
12.	Nitrofen			
13.	Paraquatdimethyl sulphate			
14.	Pentachloronitrobenzene			
15.	Pentachlorophenol			
16.	Phenyl mercury acetate			
17.	Sodium methane arsonate			
18.	Tetradifon			
19.	Toxafen			
20.	Aldicarb			
21.	Chlorobenzilate			

22. Dieldrine
23. Maleic hydrazide
24. Ethylene dibromide
25. TCA (Trichloro acetic acid)
26. Metoxuron
27. Chlorofenvinphos

(2) List of pesticides with refused registration/ pesticides restricted for use in India

Sr. No.	Refused registration	Restricted for use
1.	Calcium arsonate	Aluminium phosphide
2.	EPM	DDT
3.	Azinphos methyl	Lindane
4.	Lead arsonate	Methyl bromide
5.	Mevinphos (Phosdrin)	Methyl parathion
6.	2, 4, 5–T	Sodium Ccanide
7.	Carbophenothion	Methoxyethyl mercuric chloride (MEMC)
8.	Vamidothion	Monocrotophos
9.	Mephosfolan	Endosulfan
10.	Azinphos ethyl	Fenitrothion
11.	Binapacryl	Diazinon
12.	Dicrotophos	Fenthion
13.	Thiodemeton / Disulfoton	Dazomet
14.	Fentinacetate	
15.	Fentinhydroxide	
16.	Chinomethionate (Morestan)	
17.	Ammonium sulphamate	
18.	Leptophos (Phosvel)	

Appendix 6. Toxicity label of pesticides

Label	Name	Level of toxicity	Oral lethal dose mg per kg body weight of test animal	Listed chemicals
	Red label	Extremely toxic	1–50	Monocrotophos 36SL, Ethofenprox 10EC, Methyl parathion 50EC, Oxydemiton methyl 25EC, Zinc phosphide, Ethyl mercury acetate and others
	Yellow label	Highly toxic	51–500	Acetamiprid 20SF, Alpamethrin10EC, Chlorophyriphos 20 EC, Carbosulfan 25 EC, Cypermethrin 10 EC &25 EC, Dimethoate 30EC, Dacamethrin 2.8 EC, Deltamethrin 2.5 WP, Diazinon 20EC, Dichlorvos 76 EC, Endosulfan 35 EC, Fenvelarate 20EC, Fenpropathrin 10 EC & 30EC, Fenitrothion 50 EC&40 WP, Fenthion 82.5 EC, Fenobucarp 5EC, Fipronil 5 SC, Imidachloropid 17.8 EC, Lamdacyhalothrin 2.5 EC, Profenophos 50EC, Quinalphos 25EC, Triazophos 40 EC etc.
	Blue label	Moderately toxic	501–5, 000	Bifenthion 10EC, Carboxin 75 WP, Cartap hydrochloride 50 SP, Dicofol 18.5 EC, Flufenoxuron 10 EC, Fenazoquin 10EC, Isoprothiolane 40EC, Kitazin 48EC, Metalaxyl 35 WS, Propineb 70 WP, Propiconazol 25 EC, Penconazol 10 EC, Malathion 50 EC, Milbinectin 1 EC, Novaluron 10EC, Properjit 57EC, Thiomethoxam 25 WG, Thiram 75WP, Tridemorph 80 EC
	Green label	Slightly toxic	More than 5,000	Captan 50 WP, Carbendazim 50 WP, Clothinidin 50 DG, Hexaconazole 5 EC, Kasugamycin 3 SL, Mancozeb75WP, Sulphur 80WP, Thiophanatemethyl 70 WP etc.

INDEX